T0134051

COMMERCIALIZING NANOMEDICINE

COMMERCIALIZING NANOMEDICINE

Industrial Applications, Patents, and Ethics

edited by
Luca Escoffier
Mario Ganau
Julielynn Wong

Pan Stanford Publishing

Published by

Pan Stanford Publishing Pte. Ltd.
Penthouse Level, Suntec Tower 3
8 Temasek Boulevard
Singapore 038988

Email: editorial@panstanford.com
Web: www.panstanford.com

British Library Cataloguing-in-Publication Data
A catalogue record for this book is available from the British Library.

Commercializing Nanomedicine: Industrial Applications, Patents, and Ethics

Images/drawings on section pages: Courtesy of Lara Prisco, MD, MSc

ISBN 978-981-4316-14-9 (Hardcover)
ISBN 978-981-4613-14-9 (eBook)

Printed in the USA

Contents

Preface xi

SECTION I

1 Nanomedicine: The Dawn of a New Era **3**
Mario Ganau, Marco Paris, Nicola Nicassio, and Gianfranco K. Ligarotti
1.1 Facts and Forecasts 3
1.2 Aims of Nanomedicine 6
1.2.1 Improving Diagnostic Tools 7
1.2.2 Improving Drug Delivery 9
1.2.3 Improving Surgical Strategies 12
1.3 Organizations and Open Projects 14

2 Diagnostic Challenges of Nanomedicine **21**
Mario Ganau, Alessandro Bosco, Pietro Parisse, and Loredana Casalis
2.1 Beyond Conventional Diagnostics 21
2.2 Miniaturization and Functionalization Techniques 23
2.2.1 Conventional Photolithography and Beyond 24
2.2.2 Micro- and Nanocontact Printing 27
2.2.3 Dip-Pen Nanolithography 29
2.2.4 AFM Nanografting 30
2.3 Evolution of Biosensors for Diagnostics 33
2.3.1 Optical Readout 34
2.3.2 Radiolabeled Readout 36
2.3.3 Mass Detection Readout 37
2.3.4 Mechanical-Sensing Readout 39
2.3.5 DNA-Directed Immobilization 40

2.4 In vitro Diagnostics 41
2.4.1 Trends in Single-Cell DNA Barcode Analysis 42
2.4.2 Trends in Carbon Nanotubes and Their Use in Microfluidic Devices 44
2.5 In vivo Diagnostics 46
2.5.1 Implantable Devices 46
2.5.2 Improvements in Conventional Radiology 48
2.6 Future Trends 49

3 Surgery in the Realm of Nanometers **59**
Mario Ganau, Roberto Israel Foroni, and Rossano Ambu
3.1 Introduction 59
3.2 Potential Implications in the Field of Nanotechnology and Regenerative Medicine 60
3.2.1 Nanomaterials 60
3.2.2 Next Generation's Prostheses 62
3.2.3 Gene Therapy and Surgical Procedures 64
3.2.4 Preventing Postoperative Fibrosis and Excessive Cicatrization 65
3.3 Cellular and Subcellular Surgical Procedures 66
3.3.1 Design Improvements of Nanosurgical Instruments 68
3.4 Nanorobotics 69
3.4.1 Miniaturized Propelling Systems 70
3.4.2 Artificial Cells 71
3.5 Brain–Machine Interfaces 73
3.6 Conclusions 75

4 Nanotherapeutics **81**
Julielynn Wong
4.1 Introduction 81
4.2 Biological, Physical, and Chemical Properties and Pharmacological Effects of Nanotherapeutics 82
4.3 Nanoformulations 85
4.3.1 Nanoemulsions 85
4.3.2 Liposomes 87
4.3.3 Micelles 88
4.3.4 Nanogels 88

4.3.5 Dendrimers 89
4.3.6 Polymer Conjugates 90
4.3.7 Gold Nanoparticles 91
4.3.8 Magnetic Nanoparticles 93
4.3.9 Improved Cancer Therapies with Nanobased Drug Carriers 93
4.3.10 Encapsulating Nanovaccines 94
4.3.11 Nanobead Vaccines 95
4.3.12 Micronanoprojection Vaccines 95
4.3.13 Combined Design Approaches for Nanovaccines 95
4.4 Commercial Advantages of Nanoformulated Agents 96
4.5 Nanomedicine Publications, Patents, Product Development, and Companies 96
4.6 Current Challenges and Priorities 97
4.7 Nanotoxicity 98
4.8 Future Trends 99

SECTION II

5 Ethics and Nanoethics 111
Mario Ganau, Lara Prisco, and Laura Ganau
5.1 Introduction 111
5.2 Learning from Our Past 112
5.3 Utopian Promises, Dystopian Fears 113
5.4 Development Regulations 114
5.5 Evolving Our Nature 115
5.6 Conclusions 116

6 Nanomedicine Policy and Regulation Schemes 119
Sarah Rouse Janosik
6.1 Overview 119
6.2 Standardization of Nanotechnology Terminology and Characterization Methodologies 120
6.2.1 Standardization of Nanomedicine Terminology 121
6.2.2 Standardization Organizations 122
6.2.2.1 ASTM International Committee E56 122
6.2.2.2 ISO/TR 12802:2010 123

6.2.3 Standardization through Patent Systems 123
6.2.4 Physicochemical Characterization of Nanomedical Compounds and Devices 124
6.3 Regulation of Nanomedicine 124
6.3.1 US Food and Drug Administration 124
6.3.2 US Environmental Protection Agency and US Consumer Product Safety Commission 125
6.3.3 European Medicines Agency 126
6.3.4 Health Canada 127
6.4 Conclusion 128

SECTION III

7 Overview of Intellectual Property Rights **135**
Wim Helwegen and Luca Escoffier
7.1 Patents 136
7.1.1 Requirements 136
7.1.1.1 Patentable subject matter 137
7.1.1.2 Novelty 138
7.1.1.3 Industrial application/utility 139
7.1.1.4 Inventive step/nonobviousness 140
7.1.1.5 Disclosure 140
7.1.2 Postgrant 142
7.1.3 Exempted Uses 143
7.2 Other forms of Intellectual Property 144
7.2.1 Utility Models 144
7.2.2 Copyright 145
7.2.2.1 Economic rights, moral rights, and other features 146
7.2.3 Trademarks 146
7.2.3.1 Requirements and characteristics 147
7.2.4 Industrial Designs 148
7.3 Trade Secrets 148

8 IP Valuation: Principles and Applications in the Nanotechnology Industry **151**
Efrat Kasznik
8.1 Overview of IP Valuation 151

8.1.1 What Is Intellectual Property Valuation? 151
8.1.1.1 Brief history of IP valuation in the United States 151
8.1.1.2 IP valuation methodologies 152
8.1.1.3 Types of intangible assets 153
8.1.2 IP Valuation Standards in the United States 154
8.1.2.1 IP valuation landscape in the United States 154
8.1.2.2 IP valuation for litigation damages 155
8.1.2.3 IP valuation for financial reporting 157
8.1.2.4 IP valuation for tax reporting 159
8.1.3 IP Valuation Circumstances in Europe 160
8.1.3.1 Litigation damages 160
8.1.3.2 Financial reporting 161
8.1.3.3 Tax reporting 161
8.2 The Application of IP Valuation in the Nanotechnology Industry 162
8.2.1 IP Valuation vs. Evaluation 162
8.2.2 Valuing an IP portfolio in the Nanotechnology Industry 163
8.2.2.1 Patenting along the value chain 163
8.2.2.2 Technology transfer from university to industry 164
8.2.2.3 Mitigating litigation risk 165

9 Commercialization, Valuation, and Evaluation of Nanotech Innovations 169
Luca Escoffier
9.1 The Commercialization of Nanotechnology Innovations 169
9.2 Cost, Price, and Value: Patent Valuation vs. Patent Evaluation 176
9.2.1 Patent Valuation 178
9.2.1.1 Cost approach 180
9.2.1.2 Market approach 181
9.2.1.3 Income approach 182
9.2.2 Patent Evaluation 184
9.2.2.1 Market 184

9.2.2.2 Regulatory and legal issues 185
9.2.2.3 Technology 185
9.2.2.4 Financing 186
9.3 Introducing a Novel Approach: "Present Value After Evaluation" 187
9.4 Conclusions 204

Notes on the Contributors 213
Index 219

Preface

Following two decades of tremendous advancements in biomedical research, nanomedicine has now spread from labs to hospital wards and is already starting to redesign the state of the art in clinical practice. Many of the technological advancements achieved to date are offering new unexpected benefits to millions of patients, both in terms of hyperearly diagnosis as well as in terms of regenerative medicine; yet those are excellent premises for an even larger impact on the future of the entire humankind.

Many healthcare providers, business leaders, and lawyers are well aware that, in the coming years, medicine will be further implemented by a wide range of novel materials with bioactive and therapeutic properties, rare diseases will be defeated by cell therapy, and tissue engineering will overcome the limits of organ transplantation. Their needs are together similar and multifaceted: some of these professionals are still looking for a comprehensive guide to understand in detail this amazing revolution, some are trying to address new challenges in clinical research, those that will hopefully lead to the next Nobel Prize, and others are still wondering how to prevent improper use of these innovative technologies by patients.

To offer useful answers to the quests of such a mindful audience, and eventually to seize the entrepreneurial chances offered by the introduction of nanotechnology into fields as diverse as drug development, tissue engineering, or personalized medicine, an easy and crystal-clear guide is mandatory.

Accordingly, this book intends to cover the latest trends in nanomedicine, while providing the readers with the need-to-know ethical and commercial factors that are accompanying this epoch-changing shift in the way medicine is conceived and practiced across the globe. With this in mind, we have designed each chapter in order to offer new interesting insights to a broad, interested public, represented by scientists, surgeons, or lawyers who are meant to play a pivotal role in raising the attention to nanospecific issues, and the entire spectrum of related opportunities, in their own communities.

To provide our readers with the best-possible information, we relied upon the expertise of a great and diverse team of authors.

Also, we wish to extend our deepest gratitude to them for sharing their expertise and for their commitment and diligence during the entire process. We are also very grateful to Stanford Chong, the publisher of this work, and to his editorial team for having made the realization of a book with so many authors a smooth and enjoyable experience.

Luca Escoffier
Mario Ganau
Julielynn Wong
Tokyo, Washington, and Toronto
July 2015

Section I

Chapter 1

Nanomedicine: The Dawn of a New Era

Mario Ganau,[a] Marco Paris,[b] Nicola Nicassio,[c] and Gianfranco K. Ligarotti[d]

[a] *Department of Surgical Science & Graduate School of Biomedical Engineering, University of Cagliari, Italy*
[b] *Guy's and St Thomas NHS Foundation Trust, London, UK*
[c] *King's College Hospital NHS Foundation Trust, London, UK*
[d] *Institute of Aerospace Medicine, Milan, Italy*
mario.ganau@singularityu.org

1.1 Facts and Forecasts

In 1950 one in twelve Americans was over 60 years old; according to the Pew Research Center, by 2050 this ratio is expected to rise to one in five. Other developed countries are showing similar trends, with an associated increase in the prevalence of chronic pathologies, as such conditions are obviously age related [1]. In the U.S. only those diseases are already responsible for two-thirds of all direct healthcare costs. Those concerns do not affect richer economies only, as even developing countries are facing similar trends, if not in terms of an aging population, then at least in terms of an increase in the prevalence of such chronic conditions, in addition to the illnesses they are accustomed to treating.

Commercializing Nanomedicine: Industrial Applications, Patents, and Ethics
Edited by Luca Escoffier, Mario Ganau, and Julielynn Wong

ISBN 978-981-4316-14-9 (Hardcover), 978-981-4613-14-9 (eBook)
www.panstanford.com

Table 1.1 Forecast of healthcare spending in selected OECD countries according to the current trend

OECD countries	2030	2050	2070
U.S.	24.9	36.7	65.6
U.K.	13.5	19.9	35.6
Switzerland	18.8	27.8	40.8
France	18.0	26.6	47.6
Italy	14.5	21.3	38.2
Germany	17.4	25.6	45.9
Canada	15.9	23.5	42.0

Recently, a McKinsey Global Institute report projected that healthcare spending in the U.S. could reach 30% of the GDP by 2030 and suggested that if the current trend continues healthcare systems will certainly keep on draining an ever-growing and unsustainable proportion of the nation's wealth (see Table 1.1), possibly reaching 36.7% and 65.6% of the GDP by 2050 and 2070, respectively [2, 3].

Although difficult to figure out nowadays, the majority of these forecasts also propose an epochal shift in this trend very soon, describing the planet earth, yet by 2050, as a world where the socioeconomic impact due to the increased life expectancy of a nine-billion-people population will be definitely unburdened by technology. Such forecasts advocate the exponential growth rate, over the past two decades, of new technological discoveries that found an immediate application out of the labs in real life as the main reason for these more optimistic projections.

Indeed, delivering effective, safe, and affordable healthcare systems to such a growing population requires the application of a new paradigm to medicine, one based on high-throughput technologies replacing the existing diagnostic and therapeutic methods. To this regard, the innate dream of manipulating matter at the atomic and molecular scales has already been virtually realized in all branches of nanotechnology, including optical systems, electronics, chemistry, and engineering.

The worldwide market for nanoscale devices and molecular modeling has grown up 28%/year from 2002 to 2007, with a

35%/year growth rate in revenues, specifically from biomedical nanoscale devices [4]. To sustain this market the annual US federal funding for R&D in nanotechnology, which in 2002 barely exceeded $500 million, almost doubled within a couple of years. On the other side of the Atlantic Ocean, in the same time frame (2003–2006), the European Union had set aside €1.3 billion, while the worldwide investment had reached $3 billion in 2003 only [5, 6]. The National Institutes of Health (NIH) roadmap nanomedicine initiative first released in 2003 envisioned that this cutting-edge area of research could begin yielding benefits as early as in the 2010–2020 decade [7]. To support this endeavor a handful of nanomedicine centers staffed by interdisciplinary scientific crews, including biologists, physicians, mathematicians, engineers, and computer scientists, have been created. The aim of these crews was not only to gather extensive information about how molecular machines are built but also to create a new lexicon, a new kind of vocabulary to define biological parts and processes in engineering terms. Nanomedicine, which literally means applying the basic concepts of nanotechnology to medicine, exploits the improved and often novel physical, chemical, and biological properties of materials at the nanometer scale, and certainly it appears as the next big thing in this continuum of scientific evolution. Nanomedicine has the potential to enable early detection and prevention and to essentially improve diagnosis, treatment, and follow-up of diseases; all these improvements are especially due to the unlimited applications in such a complex translational field and based on the expected solutions to its unmet needs.

Although nanomedicine is still in its infancy intelligent surface coatings, biosensors made of smart nanoscale materials and drug nanospheres are just a few examples of the bright potentialities offered by this multidisciplinary area. Over the next 5–10 years remarkable advances in molecular medicine will open the doors to engineered organisms or even microbiological robots. In the longer term, perhaps 10–20 years from today, the earliest molecular machine systems and nanorobots may join the medical armamentarium, finally giving physicians the most potent tools imaginable to conquer not only human diseases but also more physiological aging processes [8]. Of course, nanomedicine will face a number of challenges along this path: besides the obvious scientific ones, the more complex will be probably due to its impact in our society in terms of ethical standards and legal regulations. In this book we will discuss these specific topics, which are to be considered strategic in the forthcoming development and dissemination of nanotechnology-derived medical solutions.

1.2 Aims of Nanomedicine

The potential applications of nanotechnology for diagnosis, prevention, and treatment of diseases are currently very broad. Nanomedicine relies on (1) a detailed biology and pathophysiology knowledge of diseases to enable efficient targeting and therapy, (2) thorough awareness of physical properties to manipulate matter at the nanoscale and design new diagnostic/therapeutic systems, and (3) chemical knowledge to finally provide smart modification of their surfaces and improve their biocompatibility [9]. Most of those advantages rely on the miniaturization power of fundamental techniques. For instance, the resolution of nanotechnology platforms is nowadays 3 orders of magnitude lower than at the time when the biochip was first developed; as a result the amount of information that can be put on it has increased by a factor of 10^6–10^8, thus demonstrating the powerful capabilities of nanoscaling in biomedical applications [9, 10].

The concepts expressed in Table 1.2 show how techniques gathered from basic sciences are indeed opening the path toward a more-than-needed personalized medicine. In fact, differently from

Table 1.2 Medicine versus nanomedicine: a shift in the problem-solving approach

Conventional approaches to medical problems	Nanotechnology-derived solutions in the medical field
Clinical observation driven	Laboratory assay driven
Expensive process of innovation	Cheap replication of basic science discoveries
Long design and patterning	Rapid synthesis and self-assembly
Less scalable materials	More scalable materials

the existing standard of care, the mass application of new diagnostic screening and tools in nanomedicine will allow for their fast, convenient, and inexpensive integration in the clinical practice, enabling a concrete, previously unseen, patient-centered medicine. For instance, while the common approach of conventional medicine is to remove en bloc diseased tissues, nanomedicine is currently attempting to use sophisticated approaches to either repair specific cells, one cell at a time, or detect them as soon as treatments are more effective in order to restore the organ's function.

Since an understanding of the many practical applications of nanomedicine requires a comprehensive, systematic approach, and the field is currently expanding at a very fast pace, we will try to mention the most innovative aspects in the present introduction and then expand them in the following chapters.

1.2.1 *Improving Diagnostic Tools*

Biochip analysis on a multicell level is now well accepted in clinical diagnostics in several fields: for example, expression chips for the follow-up of oral bacterial infections have made such a significant progress to be now used as point-of-care diagnostics. One goal of medical diagnostics is to progress toward the analysis of single cells. The added value of this approach becomes clear when it is taken into consideration that large amounts of primary cells are usually mixtures of either different cell types or healthy and tumoral cells, thus making the acquisition of statistically significant results extremely difficult. The ability to describe one specific cell type leads to a better understanding of the role of these building blocks in tissues and organisms and thus their function in cell–cell interaction, cell differentiation, etc. To this regard, several techniques have been developed to allow for the study of single cells; just to mention a few, cloning rings, laser microdissection, and live-cell catapulting are now available for the isolation of single adherent cells, while magnetic sorting, column chromatography, and various microfluidic approaches are commonly used for nonadherent cells [9, 11].

Moreover, the modification of biochip surfaces by nanotechnological methods offers the possibility of designing ever smaller probes for the analysis of the RNA retrieved from a single cell.

The concept that cells' behavior under pathologic conditions can influence their microenvironment by expressing protein fragments, which eventually make their way into the circulation, or secreting proteins that interact with each other and directly affect pivotal metabolic pathways has been known for a while. Indeed, the activation of oncogenes is a typical feature in diseased patients, as are the specific proteins so forth expressed, which in turn alter normal cellular behavior, resulting in the disease phenotype. Only recently these processes have been at the center of investigation through mass spectrometry techniques used in combination with mathematical algorithms to diagnose disease [12, 13]. There is substantial evidence suggesting that the low-molecular-weight circulatory proteome contains information capable of detecting early-stage diseases [14–16]. Current technologies can generate a good-resolution portrait of the proteomes; nevertheless the identification of these molecules requires techniques even more sensitive than current mass spectrometry or enzyme-linked immunosorbent assay (ELISA) techniques, which are already capable of detecting concentrations at or below 10^{-12} mol/L. The need for increased sensitivity in current techniques limits the full exploitation of such methodologies into clinical practice [17]. On the other hand emerging nanotechnology solutions that aim to amplify and harvest these biomarkers are playing a key role in discovering and characterizing molecules for early disease detection, subclassification, and predictive capability of current proteomics modalities.

Finally, by directly probing cellular properties, controlling and intensifying their physical and chemical processes, and making possible real-time direct access to intracellular biological events, nanomedicine is applying the latest discoveries of basic research at the nanoscopic level also to transform the conventional diagnostic approaches to date used in clinical imaging. One of the key benefits of nanoparticle-based materials applied to a contrast medium, an important class of imaging aids that are used for direct visualization of organs and cells, is that they might provide better information about the extent of infiltrating tumors than state-of-the-art contrast agents.

Moreover, microscopic imaging is also expected to take advantage of novel fluorescent dyes meant to yield higher photoluminescence and photostability, therefore providing better direct visualization of cells and molecules. Supported by integrated nanoanalytical models, the realm of diagnostics at the molecular level will be principally focused on the identification of pivotal biomarkers in ever smaller biological samples. As a result, different technical developments such as 4-aminolevulinic acid for intraoperative tumor visualization, navigational systems, and intraoperative magnetic resonance (MR) imaging have been introduced into daily practice, resulting in the birth of nanoneurosurgical techniques that are meant to yield significant facilitation in achieving a higher extent of resection.

1.2.2 *Improving Drug Delivery*

To date designing nanoparticles for drug delivery purposes has been the most challenging subfield in pharmacology worldwide. Targeted and controlled drug delivery relates to the ability to administer therapeutics to a patient at the right location and time, respectively, for that individual patient's need, while avoiding drawbacks to healthy organs and tissues [18–20].

The inner properties of those delivery systems (size, charge, surface hydrophilicity, and especially the nature and density of the ligands on their surface) play a significant role in determining their circulating half-life and biodistribution. Therefore many developing approaches are underway to improve the tight regulation of such novel carriers and to overcome the current challenges represented by drug insolubility and biological barriers (i.e., the blood–brain barrier).

The expected advantages are summarized in Table 1.3.

The realm of these delivery systems is rapidly evolving. Let's consider nanovectors; their class includes nanopores, micelles, liposomes, dendrimers drug–polymer conjugates, nanoemulsions, and many others [21–24]:

- Nanopores could be considered one of the earliest therapeutically useful nanomedical devices. They consist of tiny cell-containing chambers within a single-crystalline silicon wafer. They are micromachined to present a high density of

Table 1.3 Advantages of nanodrugs

Promises of nanodrugs
Improved delivery of poorly water-soluble drugs
Targeted delivery of drugs
Improved transcytosis of drugs across tight epithelial and endothelial barriers
Effective deliver of large macromolecule drugs to intracellular sites of action
Effective codelivery for combination therapy
Real-time visualization of the pharmacologic targets by incorporating imaging modalities with therapeutic agents
Improved stability of therapeutic agents against enzymatic degradation from nucleases and proteases

uniform nanopores as small as 20 nm in diameter; these pores are large enough to allow small molecules such as insulin to be loaded and transported but are small enough to prevent the passage of immune system molecules such as immunoglobulins.

- Micelles are supramolecular constructs formed through the self-assembly of amphiphiles forming a core–shell structure in an aqueous environment. They are typically realized with polyethylene glycol (PEG) and a low-molecular-weight hydrophobic core-forming block.
- Liposomes are uni- or multilamellar vesicles with a phospholipid membrane enclosing an aqueous volume; they can vary in size, surface charge and lipid composition.
- Dendrimers are polymeric complexes comprising branches around an inner core that possesses properties similar to macromolecules. They can be functionalized with carbohydrates, peptides, silicon or other chemical groups due to their high branching. They represent an attractive drug delivery system due to the presence of a central cavity and channels between the dendrons where the drugs can be trapped [17].
- Drug–polymer conjugates. Polymer–drug conjugates were some of the first nanotherapeutic platforms to be used for drug delivery. The mechanism of drug delivery involves the simple conjugation of drugs or proteins to water-soluble

polymers. Some common polymers used are PEG, *N*-(2-hydroxypropyl)methacrylamide (HPMA) copolymers, and polysaccharides. Those conjugates offer several advantages: reduced immunogenicity, increased circulation time and reduced passive targeting. This third advantage is due to the enhanced permeability and retention (EPR) effect, a phenomenon that occurs in solid tumors characterized by defective vascular architecture and compromised lymphatic systems due to pathologic angiogenetic processes [25].
- Nanoemulsions are dispersions of oil and water in which the dispersed phase forms nanosized droplets that are stabilized with a surface-active film comprising surfactants. While attractive for drug delivery due to their simple formation, thermodynamic stability, and optical transparency, their structure largely affects the drug release rate.

Due to their small size, nanodrugs are studied to penetrate deep into tissue through small capillaries or fenestration in the epithelial lining; one of their purposes is to be actively taken up by cells. The former strategy, called passive targeting, allows for more efficient delivery of the therapeutic agent to the desired target site and can be utilized in tumors or inflamed tissue that are characterized by a leaky vasculature through which nanoparticles can extravasate to the diseased site. On the other hand the latter strategy, called active targeting, exploits the potential to further tune the nanodrugs through specific ligand conjugation for various disease targets [23, 24].

For passive targeting to be effective, the nanoparticles must be studied to circulate systemically for extended periods of time; the nanoparticles therefore must avoid opsonization by the reticuloendothelial system (RES), and a common method used to circumvent this process is the conjugation of particles to the hydrophilic and low-immunogenic PEG polymer. Concerning active targeting, specific molecular moieties such as epitopes or receptors overexpressed in certain diseases can be exploited to enhance the therapeutic effects of nanoparticle drug delivery in localized diseases [24]. As described in the previous paragraph active targeting can also be used for diagnostic imaging purposes.

Targeting surface receptors is particularly advantageous whenever the delivery system is incapable of being effectively taken up by cells and needs assistance through fusion, endocytosis, or other processes to pair with their cellular target.

Finally, the greatest advantage that nanodrugs hold is the possibility to be designed as "smart" systems, meaning that they can be responsive to external stimuli. For example, nanoparticles can be designed to dissolve only under specific conditions on in certain environments, thus releasing their contents in a very spatially and timely localized manner [22]. Despite the fact that this research field has been opened by liposomes some time ago, and the work is ongoing for almost 20 years, the majority of currently approved nanotherapeutics lack the characteristics of active targeting or triggered drug delivery: currently approved products for cancer therapy accumulate in the tumor tissue, thanks to the EPR effect, and culminate in the release of their payload.

Therefore there is still a long way to go to fully overcome the challenges of long-term treatment, narrow therapeutic windows, complex dosing schedules, combination therapies, individualized dosing regimens, and unstable active ingredients [26]. Ideally, implantable devices, tailored for specific applications by the release mechanisms, reservoir geometry, and drug formulation, could soon provide people with relief from self-medication and/or recurrent doctor visits by protecting the drug from the body until administration time and allowing either continuous or pulsatile delivery of both liquid and solid drug formulations [26].

1.2.3 *Improving Surgical Strategies*

The emergence of nanotechnology offers a new tool set for the discovery, engineering, and manufacturing of nanopatterned surfaces and nanostructured scaffolds for implantable devices. Nevertheless, nanomedicine is expanding its boarders into surgery not only by redesigning the way prosthesis are made but also by introducing new tools for techniques as far as tissue engineering or bleeding control.

To date, nanotechnology offers novel and improved solutions for the localized release of the biomolecules and growth factors that are needed in any tissue engineering approach and therefore overcomes many downfalls that micron-structured implants face. To achieve optimal tissue growth, the natural extracellular environment must be mimicked for the necessary cell adhesion, mobility, and differentiation to occur [27]. To promote adequate cell differentiation and proliferation, certain features at the nanoscale are essential. Synthetic polymers are addressing the quest for tissue engineering scaffolds; they are capable of serving as bulk mechanical and structural platforms, as well as enabling the molecular interactions within the cells that are necessary to induce tissue healing. Most of those synthetic polymers are nontoxic, consistently available, inexpensive to create, and easy to alter [28].

However, they often lack the ability to create biological cues as natural polymers do to induce a desired cell response [29]. For this reason the cells rely on several topographical and physiochemical signals. These signals can be provided either by the proteins contained in the extracellular matrix or by the growth factors that bind to the receptors present on the cell surface. The cells determine their behavior through this interaction, adjust their response to the environment, and regulate their terminal differentiation [27]. As the cells move over a natural matrix or an artificial scaffold, they sense the presence of grooves and ridges through the extension and retraction of filopodia [28]. In response to these phenomena, the relevance of chemical modifications and physical features at the nanoscale proves crucial in the development of the ideal scaffold for the repair and growth of tissue.

Both tissue engineering and prosthesis integration are hugely affected by concerns related to infection, chronic inflammation, and poor binding with the surrounding tissue. As a means to address these issues, nanoscale features have been implemented to provide enhanced biointegration [30]. Natural tissues contain various nanometer features because of the presence of collagen fibrils and other proteins that are less than 100 nm in one dimension [30]. The nanometer-scaled surface structures enhance cellular response through mimicking natural tissue. Due to the tunability and adaptability of the manufacturing processes, several different scaffold types can be obtained and ideally optimized for the particular needs and requirements of the individual patient or application. Currently, nanomaterials have been proven to assist in the restoration of several tissues and organs.

Implanted devices within the body either fulfill a structural requirement, such as a bone replacement, or are implanted into the bloodstream, such as stents and possibly sensors in the future. To promote proper integration of the implants within the body, a nanotexture on the surface of the implants is essential so that the cell responds by excreting extracellular matrix molecules, humanizing the implant surface. Long-term implants with rubbing surfaces, such as joint replacements, generate particles as a consequence of wear. Various current materials produce microparticles, which lead to inflammation, while newly developed nanomaterials generate nanoparticles, which are less detrimental for the body and are much easier to be excreted and cleared from the system [17].

Hemostasis is a major problem in surgical procedures and after major trauma: From accidents to the operating room to the battlefield, uncontrolled loss of blood often means loss of life; therefore the search for devices able to stop bleeding anywhere in the body within a few minutes has always been very active. There are few effective methods to stop bleeding without causing secondary damage. Ellis-Behnke et al. used a self-assembling peptide that establishes a nanofiber barrier to achieve complete hemostasis immediately when applied directly to a wound in the brain, spinal cord, femoral artery, liver, or skin of mammals. This approach showed to be effective in stopping bleeding without the use of pressure, cauterization, vasoconstriction, coagulation, or cross-linked adhesives within a few seconds ($\sim$15 s). The self-assembling solution, which is nontoxic and nonimmunogenic, and the breakdown products, amino acids, could fundamentally change how much blood is needed during surgery of the future [31].

1.3 Organizations and Open Projects

As shown in this chapter nanotechnology will truly transform our future, especially in terms of medicine. A number of governmental organizations, academic institutions, centers of excellence, and think tanks have therefore been created to speed up the development of nanomedicine by promoting the beneficial uses of these revolutionary technologies and reduce misuse and accidents potentially associated with them. For their useful role in the scientific community and in society as a whole, and for the useful resources that they can provide to any interested reader, some of them will be herein presented.

Following the US initiatives already established during the past decade (i.e., the already cited NIH roadmap nanomedicine initiative) Europe has been very active in more recent years. The European Technology Platoform (ETP) Nanomedicine, an initiative led by industry and set up together with the European Commission, is addressing the application of nanotechnology to achieve breakthroughs in healthcare. The ETP supports its members in coordinating their joint research efforts and improving communication amongst the members as well as toward the European Commission and the European member-states. Noteworthy, one of the goal of the ETP will be directed toward an active contribution to the three pillars of the Framework Programme Horizon 2020, which are industrial leadership, excellence in science, and societal challenges. Along this path the Nanomed2020 Support Action project was launched in September 2012. This project, aiming at building a pertinent European nanomedicine community involving all key players necessary to define the resources, gaps, and needs for development and implementation of nanomedical research into marketable innovations to be used by doctors for the benefit of patients, is a great opportunity to reinforce the collaboration with the clinical community, initiate or implement public–private partnerships, and create novel infrastructures and innovative funding programs [32].

As shown in Table 1.4, several other forums exist and are very active worldwide.

Scientific societies completely devoted to the themes of nanomedicine have started to grow in terms of affiliates and prestige since the past decade. Their role within the academic community is certainly fostering this cutting-edge research by providing new resources, tools, and networks. For instance, the American Society for Nanomedicine (ASNM), a professional academic and medical

Table 1.4 List of national and international institutions with specific focus in nanomedicine

Association institution	Web resources
NanoBusiness	http://www.nanobusiness.org/Alliance
Canadian NanoBusiness Association	http://www.nanobusiness.ca
Asia Pacific Nanotechnology Forum	http://www.apnf.org/content.php?content.1
The Royal Society of the UK	http://www.royalsoc.ac.uk/landing.asp?id =1210
Northern California Nanotechnology Intiative	http://www.ncnano.org/
International Association of Nanotechnology (IANT)	http://www.nanotechcongress.com/index.htm
CSIRO: Australia's Commonwealth Scientific and Industrial Research Organisation—Nano Science Network	http://www.csiro.au/index.asp?type=blank& id=Nanotechnology_Home
Nano Science and Technology Institute (NSTI)	http://www.nsti.org/about/

society dedicated to the advancement of nanomedicine, aims to promote research related to all aspects of nanotechnology, engineering, biochemistry, molecular biology, and medicine and provide a forum for presentation of basic clinical and population-based research. Its goals also include educating all professionals interested in this emerging field through scientific meetings and publications to enable them to develop nanotechnology and biology research to treat concomitant medical conditions more effectively; encouraging primary and secondary preventive measures and technology to reduce the incidences of various diseases; facilitating the establishment of programs and policies that will better serve early diagnosis and early treatment for patients; and promoting and facilitating the formal training of physicians, scientists, engineers, molecular biologists, statisticians, and other members and allied healthcare providers in the fields of nanotechnology, nanobiology, and nanomedicine research [33].

But these societies are also looking at a broader spectrum of beneficiaries. To this regard the British Society for Nanomedicine offers a website to orient both professionals and passionate learners to a handful of information. The website is organized into three sections:

- The **Know** section has resources linking to scientific papers, up and coming conferences and regulatory authorities to help target the global nanomedicine research needs.
- The **Explore** section provides information about nanomedicine, summaries of scientific articles, and nanonews for people without a background knowledge of science.
- The **Learn** section provides learning aids and classroom activities relating to the principles of nanotechnology as well as short videos to help with explanations [34].

Finally, while several nongovernmental international institutions have focused on issues related to the application of nanotechnology to medical issues, only a few are permanently devoted to those themes: among them one of the most noticeable is the Foresight Institute. This leading think tank and public interest organization, founded in 1986, with the mission, on the one hand, to discover and promote the upsides of and, on the other hand, to prevent the dangers of nanotechnology, artificial intelligence, biotech, and similar life-changing transformative future technologies, still represents a solid benchmark in its field.

Acknowledgments

This work has been written in the framework of the first author's PhD studies in biomedical engineering at the University of Cagliari, Italy. The deepest thanks go to Prof. Giacomo Cao, Prof. Rossano Ambu, Prof. Gavino Faa, and Prof. Amedeo Columbano.

References

1. http://www.pewhispanic.org/2008/02/11/us-population-projections-2005-2050/
2. http://www.mckinsey.com/insights/health_systems_and_services/accounting_for_the_cost_of_us_health_care
3. http://www.oecd.org/eco/outlook/2085481.pdf
4. http://www.nano.gov/2002budget.html

5. http://www.aaas.org/spp/rd/04pch25/htm
6. Editorial. (2003) Nanomedicine: grounds for optimism and a call for papers. *Lancet* **362**, 673.
7. http://www.nihroadmap.nih.gov/nanomedicine/index.asp.
8. Freitas, R. A. (2005) Current status of nanomedicine and medical nanorobotics. *J Comput Theor Nanosci* **2**, 1–25.
9. Riehemann, K., Schneider, S. W., Luger, T. A., Godin, B., Ferrari, M., Fuchs, H. (2009) Nanomedicine-challenge and perspectives. *Angew Chem, Int Ed Engl* **48**(5), 872–897.
10. Fahlbusch, St., Mazerolle, S., Breguet, J.-M., Steinecker, A., Agnus, J., Pérez, R., Michler, J. (2005) Nanomanipulation in a scanning electron microscope. *J Mater Proc Technol* **167**(2–3), 371–382.
11. Jain, K. K. (2003) Nanodiagnostics: application of nanotechnology in molecular diagnostics. *Expert Rev Mol Diagn* **3**(2), 153–161.
12. Diamandis, EP. (2004) Mass spectrometry as a diagnostic and a cancer biomarker discovery tool: opportunities and potential limitations. *Mol Cell Proteomics* 3(4), 367–378.
13. Liotta, L. A., Ferrari, M., Petricoin, E. (2003) Clinical proteomics: written in blood. *Nature* **425**(6961), 905.
14. Calvo, K. R., Liotta, L. A., Petricoin, E. F. (2005) Clinical proteomics: from biomarker discovery and cell signaling profiles to individualized personal therapy. *Biosci Rep* **25**(1–2), 107–125.
15. Hu, Y., Bouamrani, A., Tasciotti, E., Li, L., Liu, X., Ferrari, M. (2010) Tailoring of the nanotexture of mesoporous silica films and their functionalized derivatives for selectively harvesting low molecular weight protein. *ACS Nano* **4**(1), 439–451.
16. Bouamrani, A., Hu, Y., Tasciotti, E., Li, L., Chiappini, C., Liu, X., Ferrari, M. (2010) Mesoporous silica chips for selective enrichment and stabilization of low molecular weight proteome. *Proteomics* **10**(3), 496–505.
17. Sakamoto, J. H., van de Ven, A. L., Godin, B., Blanco, E., Serda, R. E., Grattoni, A., Ziemys, A., Bouamrani, A., Hu, T., Ranganathan, S. I., DeRosa, E., Martinez, J. O., Smid, C. A., Buchanan, R. M., Lee, S.-Y., Srinivasan, S., Landry, M., Meyn, A., Tasciotti, E., Liu, X., Decuzzi, P., Ferrari, M. (2010) Enabling individualized therapy through nanotechnology [review]. *Pharmacol Res* **62**(2), 57–89.
18. Emerich, D. F. (2005) Nanomedicine–prospective therapeutic and diagnostic applications. *Expert Opin Biol Ther* (1), 1–5.

19. Lebold, T., Jung, C., Michaelis, J., Bräuchle, C. (2009) Nanostructured silica materials as drug-delivery systems for doxorubicin: single molecule and cellular studies. *Nano Lett* (8), 2877–2883.
20. Majd, S., Yusko, E. C., Billeh, Y. N., Macrae, M. X., Yang, J., Mayer, M. (2010) Applications of biological pores in nanomedicine, sensing, and nanoelectronics. *Curr Opin Biotechnol* **21**(4), 439–476.
21. Popat, K. C., Mor, G., Grimes, C. A., Desai, T. A. (2004) Surface modification of nanoporous alumina surfaces with poly(ethylene glycol). *Langmuir* **20**(19), 8035–8041.
22. Farokhzad, O. C. and Robert Langer, R. (2009) Impact of nanotechnology on drug delivery. *ACS Nano* **3**(1), 16–20.
23. Panyam, J., and Labhasetwar, V. (2003) Biodegradable nanoparticles for drug and gene delivery to cells and tissue. *Adv Drug Delivery Rev* **55**, 329–347.
24. Koo, O. M., Rubinstein, I., Onyuksel, H. (2005) Role of nanotechnology in targeted drug delivery and imaging: a concise review. *Nanomed: Nanotechnol, Biol, Med,* **1**(3), 193–212.
25. Maeda, H., Wu, J., Sawa, T., Matsumura, Y., Hori, K. (2000) Tumor vascular permeability and the EPR effect in macromolecular therapeutics: a review. *J Controlled Release* **65**(1–2), 271–284.
26. Staples, M., Daniel, K., Cima, M. J., Langer, R. (2006) Application of micro- and nano-electromechanical devices to drug delivery. *Pharm Res* **23**(5), 847–863.
27. Engel, E., Michiardi, A., Navarro, M., Lacroix, D., Planell, J. A. (2008) Nanotechnology in regenerative medicine: the materials side. *Trends Biotechnol* **26**(1), 39–47.
28. Place, E. S., George, J. H., Williams, C. K., Stevens, M. M. (2009) Synthetic polymer scaffolds for tissue engineering. *Chem Soc Rev* **38**(4), 1139–1151.
29. Harrington, D. A., Sharma, A. K., Erickson, B. A., Cheng, E. Y. (2008) Bladder tissue engineering through nanotechnology. *World J Urol* **26**(4), 315–322.
30. Chun, Y. W., Webster, T. J. (2009) The role of nanomedicine in growing tissues. *Ann Biomed Eng* **37**(10), 2034–2047.
31. Ellis-Behnke, R. G., Liang, Y. X., Tay, D. K., Kau, P. W., Schneider, G. E., Zhang, S., Wu, W., So, K. F. (2006 Oct 12) Nano hemostat solution: immediate hemostasis at the nanoscale. *Nanomedicine* (4), 207–215.
32. http://www.etp-nanomedicine.eu/public

33. http://www.amsocnanomed.org/
34. http://www.britishsocietynanomedicine.org/
35. http://www.foresight.org/

Chapter 2

Diagnostic Challenges of Nanomedicine

Mario Ganau, Alessandro Bosco, Pietro Parisse, and Loredana Casalis
NanoInnovation Lab at Elettra, Synchrotron Light Source Park, Trieste, Italy
mario.ganau@singularityu.org

2.1 Beyond Conventional Diagnostics

Diagnostics play a key role in medicine for the successful prevention or efficient treatment of diseases. Taking cancer as an example of a widespread disease, and a leading cause of death in many countries (especially the industrial ones), it will be difficult to achieve a meaningful increase in the cure efficiency unless more information about the pathophysiology can be obtained and timely translated into appropriate treatments [1].

In fact, clinicians are nowadays at a crossroads where therapeutic choices are being made considering not only conventional histological diagnosis but also the latest insights from molecular biology [2]. On the basis of clinical observations, for years assumptions have been made considering that even genetically identical cells can exhibit significant functional heterogeneity; in the last years a remarkable body of laboratory evidence confirmed it, raising the

Commercializing Nanomedicine: Industrial Applications, Patents, and Ethics
Edited by Luca Escoffier, Mario Ganau, and Julielynn Wong

ISBN 978-981-4316-14-9 (Hardcover), 978-981-4613-14-9 (eBook)
www.panstanford.com

level of discussion about common pathophysiology of cancer and degenerative diseases.

The core of diagnostics consists in empowering clinicians with information coming from accurate biosensors. A biosensor is a device that combines a biological component with a physicochemical detector used for the recognition of an analyte at a micro- or nanometer scale. Currently, two main challenges need to be addressed in order to support the development of new analytical methods for detection of pivotal pathologic markers: the first one is a technological tenet and the second a purely qualitative one:

- Miniaturization is the first technical challenge; it mainly lies in reducing the quantities of biological specimens required for diagnostic assays, while scaling down the minimum amount of DNA or proteins that can be directly detected.
- The second consists in increasing the overall knowledge in the field of protein characterization and its clinical usefulness by identifying specific pathologic biomarkers and clarifying the correlation between overexpression or underexpression of certain proteins and the subsequent clinical course.

Accordingly molecular diagnostics is rapidly moving beyond genomics to proteomics, with the aim to identify those posttranslational modifications expressed under pathologic conditions [3]. The proteome and secretome by definition are dynamics and change both in physiologic and in pathologic conditions; the ultimate goal of determining them is to characterize the flow of information within the cells, through the intercellular protein circuitry that regulates the extracellular microenvironment. Indeed, the study of proteomics and molecular biomarkers already allows us to identify direct or indirect predictive factors, and soon it will hopefully determine which affected pathway could become a selective therapeutic target. Accordingly, nanotechnology-based approaches are being extensively explored to discover, identify, and quantify clinically useful cellular signatures for early detection, diagnosis, and prognosis of any pathology.

Because of their clinical importance, the generation at the nanoscale of new materials suitable for diagnostic purposes has always been central in the conversation about nanomedicine. The potentially better functionality and bioavailability of nanomaterials led to intense researches oriented toward amorphous materials, semiconductor–quantum dot structures, nanoclusters, and nanopowders. Those materials are expected to find their place not only in laboratory assays but also in various in vivo techniques such as computed tomography or magnetic resonance imaging (MRI). The increased resolution and sensitivity, enabling earlier diagnosis of diseases, are expected to lead to cheaper clinical applications also when applied to site-specific drug delivery. Accordingly this chapter will cover both the areas of in vitro and in vivo diagnostics and will create the basis to further understanding their present and future clinical applications.

2.2 Miniaturization and Functionalization Techniques

The aim of nanodiagnostics is evident even to the common public: its ultimate goal is to provide a noninvasive, early, and accurate detection of the biological disease markers in a process of routine screening. Not only, miniaturization allows for the manufacture of portable, hand-held, implantable, or even injectable devices. In addition, as a result of their minute size, these devices need less sample or reagent for analysis or operation, saving money and time. Moreover, where materials and processes are inhibited by lengthy diffusion time, miniaturization provides a mechanism for abbreviating the latter. A notable example where these microdevices allow for significant advantages over traditional technologies is in medical care. For instance, harnessing the ability to precisely and reproducibly actuate fluids and manipulate bioparticles such as DNA, cells, and molecules at the microscale, microfluidics offers opportunities that are currently revolutionizing chemical and biological analysis by replicating laboratory bench-top technology on a miniature chip-scale device, thus allowing assays to be carried out at a fraction of the time and cost, while affording portability and field-use capability. Emerging from a decade of research and development in microfluidic technology are a wide range of promising laboratory and consumer biotechnological applications from microscale genetic and proteomic analysis kits, cell culture and manipulation platforms, biosensors, and pathogen detection systems. To date in vitro analyses are used to monitor the level of inflammatory, viral, or oncological markers, to monitor the blood availability of prodrugs or their active forms, and more in general to predict the behavior of cells under various exogenous stimuli. Furthermore, point-of-care diagnostic testing, which allows us to test directly at the patient's bedside, permits physicians to diagnose a patient's conditions more rapidly than conventional lab-based testing. By using these devices to reduce the time to diagnoses, the physician is able to make better patient management decisions, leading to improved patient outcomes and reduce the overall cost of care. Advances in microelectronics and biosensor tools have been instrumental in facilitating the development of these diagnostic devices.

Various platforms have been developed to allow for the simultaneous real-time evaluation of a broad range of disease markers by noninvasive techniques. Among them two classes of microtechnological devices developed since the early 1990s, such as microarray DNA chip and microfluidics systems for lab-on-a-chip diagnostics, have found their full application following further miniaturization at the nanoscale. Several techniques from the field of nanotechnology are nowadays available for the miniaturization and biofunctionalization of diagnostic surfaces which can find a place in the screening armamentarium for molecular analyses. Indeed, many of them appear particularly suitable for high-sensitivity determination of panels of biomarkers. The main purpose of any technique for fabrication and biopatterning of surfaces is to create very small structures (at the micro- and nanometer scale) in a resist that can subsequently be transferred to the substrate material, often by etching. Accordingly, in the following sections the basic concepts regarding photolithography, microcontact printing, dip-pen nanolithography (DPN), and atomic force microscopy (AFM) nanografting will be described.

2.2.1 *Conventional Photolithography and Beyond*

Photolithography and other lithographic techniques enable scientists to create extremely small patterns (down to a few tens of

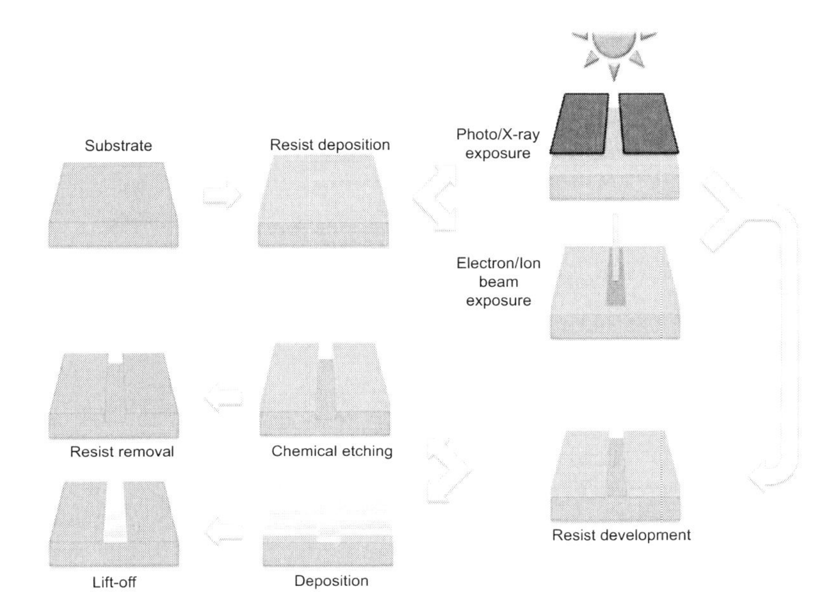

Figure 2.1 Scheme of lithographic techniques.

nanometers in size), while maintaining exact control over the shape and size of the objects created (see Fig. 2.1). Photolithography, a very cost-effective process used in microfabrication, is one of the most commonly exploited techniques for patterning of thin films: in other words it allows us to transfer a geometric pattern from a photomask to a light-sensitive chemical photoresist on the substrate [4]. Photolithography is therefore a binary process: it segregates a surface into regions that are exposed to a modification and regions that are masked from that modification. Photolithography has reached wide acceptance in the field of microfabrication because of the high resolution and variety of pattern attributes that are possible to obtain, both of which depend on the characteristics of the photomask. Nonetheless, this technique has the limitation of requiring clean room processing.

Another lithography technique is the electron-beam lithography, a practice consisting of emitting a beam of electrons in a patterned fashion across a surface covered with a film and of selectively removing either exposed or nonexposed regions of the resist. By beating the diffraction limit of light, electron-beam lithography allows us to make features in the nanometer regimen: for this it was originally developed for manufacturing semiconductor components and integrated circuits; nevertheless it found huge application also for creating nanotechnology architectures. The feature resolution limit is determined not by the beam size but by forward scattering (or effective beam broadening) in the photoresist. Noteworthy, the main drawback of electron-beam lithography is the cost associated with purchase and maintenance of the system; moreover it is a slow process, requiring long exposure times, and therefore its key limitation is due to throughput.

Interference lithography is a technique for patterning regular arrays of fine features, without the use of complex optical systems or photomasks. An interference pattern between two or more coherent light waves is set up and recorded in a recording photoresist; thus layers corresponding to the periodic intensity pattern emerge. The benefit of using interference lithography is the quick generation of dense features over a wide area without loss of focus. Hence, it is commonly used in biotechnology. A key advantage of using electrons over photons in interferometry is the much shorter wavelength for the same energy; for this electron interference lithography may be used for patterns that normally take too long for conventional electron-beam lithography to generate. One drawback of both interference and electron-beam lithography is related to secondary electrons from ionizing radiation.

Ion-beam lithography is the practice of scanning a focused beam of ions in a patterned fashion across a surface; while other methods of classical optical lithography are limited by the diffraction effect of light for nanolithography, the state of the art of focused ion-beam lithography limits the throughput for large-area batch fabrication. Focused-ion-beam lithography can also be used to pattern features directly to a substrate, without the need for a photomask, by either selective material removal or deposition. On the other hand, when compared to electron ion-beam lithography, the patterning speed offered by this technique is significantly slower because of lower achievable ion current density.

Finally, soft lithography, a set of techniques for microfabrication based on printing and molding using elastomeric stamps with the patterns of interest in base-relief, found huge application for fabricating microstructures for biological applications. In fact, soft lithography overcomes many of the shortcomings of photolithography, offering the ability to control the molecular structure of surfaces and to pattern the complex molecules relevant to biology (i.e., self-assembled monolayers [SAMs] of alkanethiolates on gold), and they provide exquisite control over surface biochemistry. As a result, since its introduction in laboratory practice soft lithography has proved very convenient, inexpensive, and rapid for fabricating channel structures appropriate for microfluidics or biochips meant to pattern and manipulate live cells [6, 7].

Nevertheless, fabrication of complex structures that contain more than two types of elements is often required for microfluidic and microelectromechanical systems and necessitates multiple applications of different lithographic techniques in which each step must be aligned spatially with previous ones. Noteworthy, alignment of patterning steps in the fabrication of organic, living, or soft structures has proven to be cumbersome for many reasons: the elastomeric stamps used in soft lithography are difficult to align over large areas, alignment of biological materials requires sterile working conditions throughout the fabrication process, and patterning onto devices that are not openly accessible (such as a sealed microfluidic device) is extremely challenging.

2.2.2 *Micro- and Nanocontact Printing*

As explained above, for a while photografting of proteins was mainly obtained from classical photolithographic techniques; in 1993 a novel approach, called microcontact printing, was introduced by Whitesides and colleagues for patterning SAMs of alkanethiols onto gold substrates [6, 7].

To date, microcontact printing is widely used for generating micropatterns of nanomaterials such as organic molecules and biomolecules over large surface areas ($>$cm^2). In the microcontact printing process, a microstructured elastomer stamp is coated with a solution of a nanomaterial and applied to a substrate of choice; then after a given period of time in conformal contact (generally

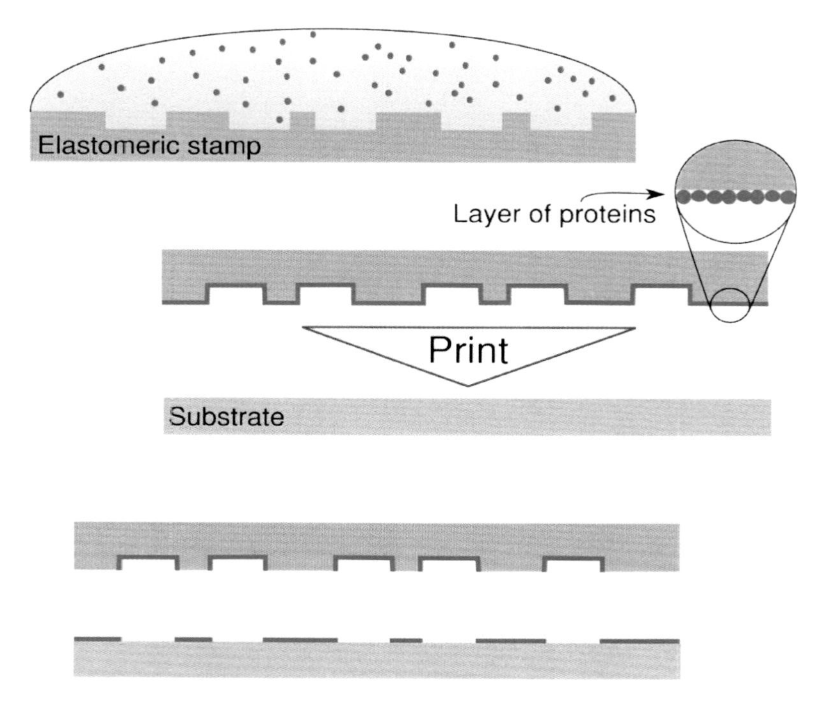

Figure 2.2 Scheme of microcontact printing: a conventionally fabricated PDMS stamp serves as the vehicle to transfer the ink of choice, in this case proteins, upon brief contact; then the transfer of those proteins occurs only at the sites of conformal contact between stamp and substrate.

overnight), the stamp is removed, leaving a replica of the stamp pattern on the substrate surface (see Fig. 2.2).

The elastomer stamps are made typically from a polymer matrix, such as poly-dimethylsiloxane (PDMS) or poly(methyl-methacrylate) (PMMA), by curing liquid prepolymers on a lithographically prepared master. The elastomeric properties of stamps made of such polymers ensure conformal contact (intimate contact) with various substrates.

As the stamps can be structured with almost any pattern, conformal contact can be achieved in many different geometrically conformed ways. Moreover, since this technique is carried out at room temperature, different biomolecules may directly be transferred in a controlled way onto a variety of substrates, while retaining their biological activity. However, the deposition of liquid samples followed by drying could be very complex, leading to ill-defined patterns, protein aggregation, and loss of biological activity [8]. Noteworthy, this problem has recently been overcome by extending these concepts to the nanoscale dimensions in a process referred to as nanocontact printing, so features as small as 40 nm can now be fabricated by this way [9]. Nanocontact printing has been achieved by decreasing the feature sizes in the PDMS stamp and diluting the nanomaterial inks, utilizing special variants of PDMS stamps, or employing new polymeric material stamps. Another important factor on obtaining high-resolution prints at the 100 nm level relates with the ink utilized; to this regard biomolecules are attractive nanocontact printing inks since their high molecular weight prevents diffusion during the printing step, resulting in high-resolution features.

As opposed to the parallel conventional photolithographic process, nanocontact printing is not diffraction limited and makes it possible to pattern surfaces with molecular-sized features. A significant advantage of nanocontact printing lithography compared to serial techniques such as DPN is that large areas can be nanopatterned rapidly. Nevertheless, multicomponent biomolecule nanopatterning is still very problematic with this technology due to the practical difficulties in accurately aligning multiple flexible stamps over a large area, while maintaining a nanoscale resolution, and thus further development is required to solve this problem.

2.2.3 *Dip-Pen Nanolithography*

While studying a process through which molecules could be transferred to a wide variety of surfaces to create stable chemically adsorbed monolayers in a high-resolution lithographic process, Mirkin and colleagues termed "DPN" a scanning probe lithography technique where an atomic force microscope tip was used to transfer alkane thiolates to a gold surface [10, 11].

DPN allows surface patterning on scales of under 100 nm, and is the nanotechnology analog of the dip pen (also called the quill pen), where the tip of an atomic force microscope cantilever acts as a "pen," which is coated with a chemical compound or mixture acting as an "ink" and put in contact with a substrate, the "paper." DPN enables direct deposition of nanoscale materials onto a substrate in a flexible manner. Recent advances have demonstrated massively parallel patterning using 2D arrays of 55,000 tips. Applications of this technology currently range through chemistry, materials science, and the life sciences and include fabrication of ultrahigh-density biological nanoarrays [12].

2.2.4 *AFM Nanografting*

Nanopatterning of surfaces for biomedical applications has been of growing interest in recent years, from both scientific and technological points of view: indeed, diverse biological and medical applications can be envisioned, such as biochips and biosensors. To this regard nanotechnology offers not only the reward of smaller dimensions with more reaction sites but also smaller test sample volumes and potentially higher sensitivity and higher-throughput screening for molecular diagnostics.

AFM is a very high-resolution type of scanning probe microscopy, consisting of a cantilever with a sharp tip (probe) at the microscope's end, which is used to scan the specimen surface, and a laser to measure any probe deflection and reflect it from the top surface of the cantilever into an array of photodiodes; the tip–sample intermolecular forces are detected as a function of the distance between the two (see Fig. 2.3).

The atomic force microscope can be operated in a number of modes, depending on the application: basically AFM imaging may be divided into static mode (i.e., contact mode) and a variety of dynamic modes (i.e., noncontact or "tapping") where the cantilever is oscillated. The stiffness of the cantilever determines the ratio between the distance moved and the force exerted by the surface; therefore this parameter is most relevant for determining the tip–sample interaction during the majority of atomic force microscope operation modes.

For instance, medium-stiffness cantilevers are well suited for a fine patterning of surfaces: if compared to other methods of nanofabrication, nanografting allows more precise control over the

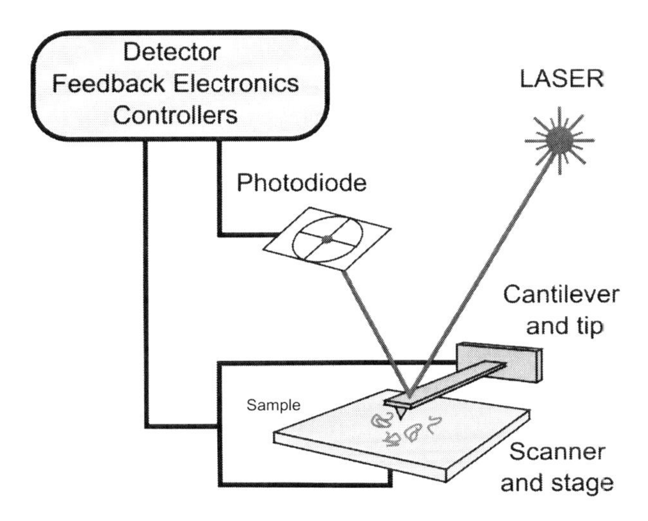

Figure 2.3 Schematic illustration of the atomic force microscope: The tip is attached to a cantilever and is raster-scanned over a surface. The cantilever deflection due to tip–surface interactions is monitored by a photodiode sensitive to laser light reflected at the tip backside; the position of the reflected beam is kept centered in the diode through feedback-controlled z-changes in the stage.

size and geometry of patterned features and their location on the surface. The technique of nanografting is usually used on SAMs and is achieved in the presence of a second replacement surfactant molecule with a greater affinity for the surface or concentration in the grafting solution than the molecule being removed by the atomic force microscope tip. Therefore, once the preformed SAM is removed from the desired area by the atomic force microscope tip, it will be replaced with a second surfactant to form a new SAM in the patterned area.

Noteworthy, some criteria need to be met: the SAM must be readily removable with the force applied by the atomic force microscope tip, but more importantly, the second surfactant must form the new SAM rapidly. It is for these reasons that thiol SAMs on gold are usually the system of choice for nanografting experiments, due to the way in which thiols rapidly form homogenous monolayers on exposed gold surfaces. This strategy may be used for the production of nanometer-sized protein patterns on gold surfaces by exploiting the affinity of biomolecules toward different SAMs [13].

AFM-based lithography methods are attractive nanoarraying techniques and have shown many potentialities in generating arrays with significantly reduced amount of capture materials, such as DNA, peptides, and antibodies. Further, these methods exploit the atomic force microscope tip (radius of curvature below 10 nm) to selectively pattern complex structures on the surface and can offer high sensitivity and resolution.

By varying fabrication parameters, such as the number of scanning lines at high tip load set in a given surface area, the speed of the atomic force microscope tip, the concentration of molecules in solution, etc., and the number of molecules released to the surface can be appropriately tuned. In the pioneering work of Mirmomtaz, our group showed that by nanografting DNA nanostructures patches of predetermined different heights could be reproducibly created [14]. Moreover recent investigations demonstrated the correlation between patch height and DNA molecules' surface density in the range of 10^{12}–3×10^{13} mol cm^{-2} [15, 16].

The great advantage of AFM patterning is that the same technique may be used for both patterning and imaging a SAM as several physical and mechanical properties can be measured all at once. The topographic height of the patches is used to infer any change at any step of the nanoassay and, concertedly with the measurement of the roughness, within and outside the patch, constitutes a unique method not only to quantify the biorecognition events but also to rule out the presence of unspecific molecular adsorption. Other advantages of this technique include

- the possible identification of molecular orientation by measuring the molecular height with high precision (order of angstroms) with respect to a supporting substrate;
- the well-defined patterning of homogeneously oriented molecules; and
- the possibility of printing multiple features in array format, where different molecules are placed selectively at different sites [17, 18].

2.3 Evolution of Biosensors for Diagnostics

Different ways to immobilize DNA molecules or specific proteins of interest have been the subject of intensive research over the past years, the strongest efforts being put into the optimization of nano- to femtomolar detection of bioanalytes onto functionalized array surfaces. Understanding the structure and function of each protein and the complexities of protein–protein interactions will be critical for shaping the most effective proteomic instruments in the future. In general, among the complex aspects characterizing the development of biosensors the most intriguing one is certainly the individual conditioning of the different elements that must be assembled: proteins, DNA, and antibodies must in fact maintain their functional conformation throughout the assay procedure, despite relevant differences in electrical charge, hydrophobicity, posttranslational modification, and folding [19]. To date a number of techniques have emerged as effective tools for the discovery of key biomarkers; nevertheless a combination of multiple techniques is mandatory to attain the goal of measuring multiple parameters in a single living cell, and only the hybridization of proteomic methods with protocols and devices for cells patterning is finally yielding to the development of arrays for few cells' proteomics.

An indispensable requirement for any biosensor is excellent specificity and sensitivity for biomarker detection; the former can be defined as the ability of the assay to rule out a condition when a specific biomarker is absent, while the latter is defined as the ability of the assay to identify a condition when it is present [20]. These clinical specificity and sensitivity parameters are closely linked to the method used for measurements and need to be high (>90%) to avoid false positive or false negative results [21]. The sensitive biological elements of a biosensor, generally represented by biologically derived materials or biomimetic components (i.e., tissues, microorganisms, enzymes, antibodies, nucleic acids, etc.), interact throughout specific binding to, or recognition of, the analyte under study. The detector element or transducer, which works in a physicochemical way (i.e., optical, piezoelectric, electrochemical, etc.), transforms the signal resulting from the interaction of the analyte with the biological element into another signal that can be more easily measured and quantified.

Detection elements play a key role in analyte recognition in biosensors: therefore detection elements with high analyte specificity and binding strength are required. While antibodies have been increasingly used as detection elements in biosensors, some key challenges remain: their immobilization on the biosensor surface and the optimal method for identifying the antigen–antibody interaction. According to the array-based optical, mass detection or radiolabeled readout, micro- and nanobiosensors can be classified as follows:

- optical read-out
- radiolabeled read-out
- mass detection read-out
- mechanical-sensing read-out
- DNA-directed immobilization (DDI)

2.3.1 *Optical Readout*

Optical biosensors account for the most known and widespread devices and protocols for detection of bioanalytes, including ELISA, which is particularly useful for a plate-based detection and quantification of substances such as peptides, proteins, antibodies, and hormones.

In ELISA, an antigen must be immobilized to a solid surface and then complexed with an antibody that is linked to an enzyme; the detection is accomplished by assessing the conjugated enzyme activity via incubation with a substrate to produce a measureable product, which generally is a detectable fluorescence signal. Therefore the most crucial element of the detection strategy is a highly specific antibody–antigen interaction.

ELISAs are typically performed in polystyrene plates, which will passively bind antibodies and proteins; having the reactants of ELISA immobilized to the microplate surface makes it easy to separate bound from nonbound material during the assay. This ability to wash away nonspecifically bound materials makes ELISA

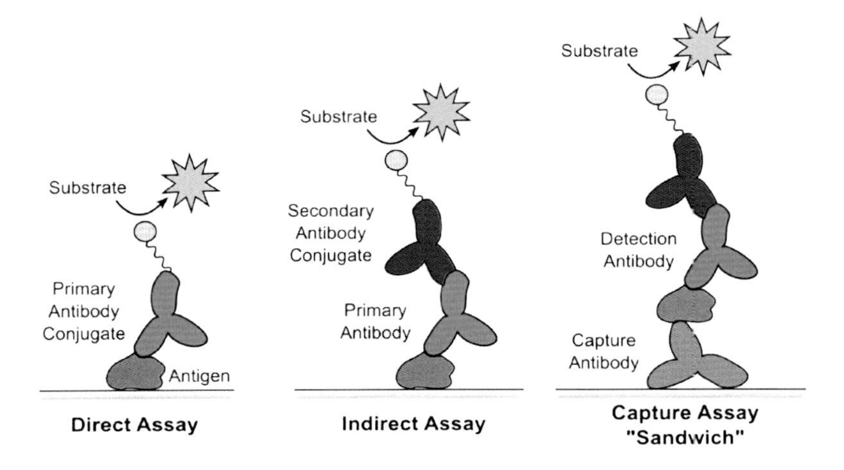

Figure 2.4 Schematic representation of the most common ELISA formats.

a powerful tool for measuring specific analytes within a crude preparation.

ELISAs can be performed with a number of modifications to the basic procedure (see Fig. 2.4): the key step, immobilization of the antigen of interest, can be accomplished by direct adsorption to the assay plate or indirectly, via a capture antibody that has been attached to the plate. The antigen is then detected either directly (labeled primary antibody) or indirectly (labeled secondary antibody). The most powerful ELISA format is the sandwich assay. This type of capture assay is called a "sandwich" assay because the analyte to be measured is bound between two primary antibodies, the capture antibody and the detection antibody. The sandwich format is often the preferred one because it is more robust and equally sensitive then direct or indirect assays [22].

ELISA is nearly always performed using 96-well or 384-well polystyrene plates and samples in solution (i.e., biological fluids, culture media, or cell lysates); however, other variants of ELISA exist:

- Enzyme-linked immunospot assay (ELISPOT) refers to ELISA-like capture and measurement of proteins secreted by cells that are plated in polyvinylidene difluoride (PVDF)-membrane-backed microplate wells. It is a sandwich assay in which the proteins are captured locally as they are secreted by the plated cells, and detection is with a precipitating substrate. ELISPOT is like a Western blot in that the result is spots on a membrane surface [23].

- In-cell ELISA is performed with cells that are plated and cultured overnight in standard microplates. After the cultured cells are fixed, they undergo a permeabilization and blocking processes, and finally target proteins are detected with antibodies. This is an indirect assay, not a sandwich assay. The secondary antibodies are either fluorescent, for direct measurement by a fluorescent plate reader or a microscope, or enzyme-conjugated, for detection with a soluble substrate using a plate reader.

2.3.2 *Radiolabeled Readout*

The measurement of radiolabels by scintillation counting has long been one of the most reliable methods for accurate, quantitative measurement in biochemical experiments [24].

Today it has been supplanted by the ELISA method, where as previously said the antigen–antibody reaction is measured using colorimetric instead of radioactive signals; however, because of their robustness, consistent results, and relatively low price per test, radiolabeled readout methods are again becoming popular [25].

The concepts of radiolabeled readout have been employed in the context of proteomics, where they offered gains in absolute sensitivity and dynamic range: for instance, multiphoton detection methodology, proposed as a tool to routinely and quantitatively detect radioactive labels on 2D gels, has several characteristics that are advantageous for functional protein detection:

- First of all, by using single-particle detectors, the sensitivity for detection of radiolabels can be improved dramatically.
- Secondly, because single-particle detectors can differentiate the particle energies produced by different decay processes, it is possible to choose combinations of radioisotopes that can be detected and quantified individually on the same 2D gel.

- Thirdly, this technology is essentially linear over 6 to 7 orders of magnitude (i.e., it is possible to accurately quantify radiolabeled proteins over a range from at least 60 zeptomoles to 60 femtomoles) [26].

In principle, the implementation of chemical radiolabeling methods could provide a 100-fold decrease in the amount of biological material needed for proteomics experiments, while reducing imaging times 10–100-fold, with total amounts of radioactivity far below legal limits [26].

Overall, the quest for ultrahigh sensitivity and quantitative precision is providing new impetus to proteomics studies: both micro- and nanoarrays hold the promise of high selectivity and sensitivity, ease of use, reasonable costs per assay, and good possibilities for future automation. Nevertheless several drawbacks still limit the diffusion of radiolabeled readout; the most important ones are certainly related to the special facility, precautions, and licensing required: Since radioactive substances are used a gamma counter is essential to measure the radiation emitted by the radionuclide, while security issues impose strict protocols for their stocking and disposal.

2.3.3 *Mass Detection Readout*

Mechanical interactions are fundamental to biology. In fact on the one hand mechanical forces of chemical origin determine motility and adhesion on the cellular scale and govern transport and affinity on the molecular scale; on the other hand biological sensing in the mechanical domain provides unique opportunities to measure forces, displacements, and mass changes from cellular and subcellular processes [27].

The advances in micro- and nanofabrication technologies have enabled the preparation of increasingly smaller mechanical transducers, so nowadays a promising family of biosensors is represented by micro- and nanomechanical systems, which are basically cantilever-like sensors: they are particularly well matched in size with molecular interactions and provide a basis for biological probes with single-molecule sensitivity, indeed [28, 29]. Recently, detection of mass in the zeptogram range and sensitivity in liquid to the fraction of nM concentration in real time has been demonstrated [30].

Despite the fact that biosensors based on nanomechanical systems have gained considerable relevance in the past decade, several theoretical and experimental studies, reporting the influence of the mass transport on antibody biosensors as a function of analyte concentration and incubation time, concluded that pushing the sensitivity to the limit of single-molecule detection may not bring the expected benefit to the overall performance [31, 32].

In fact mass transport can significantly lower the practical sensitivity of a device by reducing the number of binding events [33]. Moreover, especially at low concentration, which is typical of biomolecular experiments, the interaction between target molecules and the biosensor can play as critical a role as the chemical reaction itself in governing the binding rate [34].

In the attempt to overcome these limitation Melli et al. developed a micromechanical sensor (see Fig. 2.5) based on vertically oriented oscillating beams (or pillars), which make it possible to locate the sensitive area at the free end of the oscillators [35].

Practically, an array of such pillars (3×8 μm in plane and 15 μm in height) behaves as an array of isolated nanosized sensors embedded in a quasi-infinite analyte solution: while the top face of the pillars represents the nanosized active area, the pillars

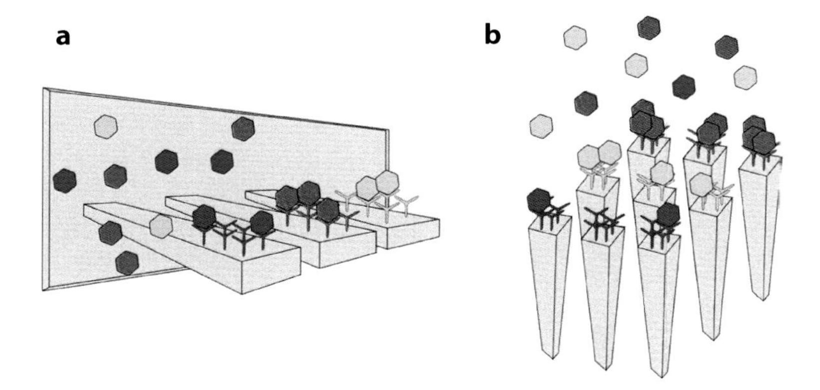

Figure 2.5 (a) Functionalized cantilevers and (b) micromechanical pillars.

themselves can be operated as mass detectors. In particular these 3D structures with dimensions comparable to the diffusion length of the target molecules have proved to increase the reaction speed by 3 orders of magnitude, while attaining improvement also in concentration sensitivity.

2.3.4 *Mechanical-Sensing Readout*

To conclude this digression on micro- and nanobiosensors it is also useful to cite that in the past decade, the quest for protein detection in smaller volumes, along with continuous efforts to monitor specific interactions between antibodies and antigens employed in an immunoassay system, has led to the development of several ELISA-like biosensors, including AFM-based arrays.

AFM is a powerful technique for investigating surfaces by visualizing their topographic characteristics: hence to obtain quality images it is critical that the atomic force microscope tip not damage the surface being scanned; soft cantilevers (<0.04 N/m) allowing us to image surfaces with very low forces are the most indicated to avoid undesirable surface modifications.

Even a very low number of target molecules can be reliably detected using height and/or compressibility measurements: starting from the works of Liu and colleagues it has been consistently demonstrated that accurate height measurements of nanopatches before and after sequence-specific hybridization of DNA oligomers allow for reliable, sensitive, and label-free detection of hybridization itself [14, 36–38].

Moreover, AFM probes may also be functionalized: in 1997, Allen et al. were the first to use AFM probes functionalized with ferritin to monitor the adhesive forces between the probe and antiferritin antibody-coated substrates [39].

Recently, Volkov et al. demonstrated that a reliable reading of the immunosignal (a suggested dimensionless combination of brush length and grafting density) can be obtained from an area as small as $\sim 3\ \mu m^2$—approximately 4 million times smaller compared to typical ELISA sensors [40]. Intriguingly, they found that AFM could reliably distinguish between having the immunosignal from either antibody and from both antibodies together, attaining a new detection limit that was impossible to obtain by using standard optical methods.

2.3.5 *DNA-Directed Immobilization*

Diagnostic immunoassays and DNA sensing are driving efforts to miniaturize biological assays and to conduct them in parallel; specifically, DNA-based arrays are becoming the leading technology for high integration and miniaturization of bioassays [41, 42].

The use of DNA microarray technology for proteomics known as DDI was introduced in 1994 by Niemeyer et al., showing that covalent DNA–streptavidin conjugates could be utilized for the reversible and site-selective immobilization of various biotinylated enzymes and antibodies [43]. Their pioneering experiments demonstrated that enzymes, such as biotinylated alkaline phosphatase, beta-galactosidase, or horseradish peroxidase, as well as antibodies, such as biotinylated anti–mouse or anti–rabbit immunoglobulins, could be coupled to the DNA–streptavidin adapters by simple, two-component incubation and that the resulting preconjugates could be exploited to hybridize complementary oligonucleotides by surface binding [44–46].

DDI proceeds with a higher immobilization efficiency than conventional immobilization techniques because of the reversible formation of the rigid, double-stranded DNA spacer between the surface and the proteins. The simultaneous immobilization of different compounds using microstructured oligonucleotide arrays as immobilization matrices demonstrates that DDI proceeds with site selectivity due to the unique specificity of Watson–Crick base pairing; moreover, it allows for a reversible functionalization of the sensor surfaces with the proteins of interest.

Since DDI technologies and DNA nanoconstruction essentially depend on similar pre-requisites, which in particular are large and uniform hybridization efficiencies, combined with low nonspecific cross reactivity between individual sequences, this microarray approach has emerged along the past years as a promising tool for chip-based immunoassays meant to multiplex antigen detection. In fact, it is well known that the self-assembly of semisynthetic DNA–protein conjugates in so-called nanoassembled monolayers

(NAMs) makes it easy to generate unlimited reproducible, configurable nanoarrays exploiting precise and reliably protein/antibodie detection methods.

DDI allows for highly economical use of antibody materials (at least 100-fold lower than the amount needed for preparing an array by direct spotting). Therefore taking into account the greater versatility and convenience of handling of the self-assembly approach, DDI proved to be an advantageous alternative to conventional techniques for generating versatile and robust protein arrays.

The DDI strategy bears the potential for relatively rapid high-sensitivity determination of limited panels of biomarkers with good precision and accuracy. Despite its potential to revolutionize protein diagnostics, the major problems in the fabrication of such antibody arrays concern the initial efforts required to reproduce homogeneously the attachment of the antibody on the DNA substrate. To this regard, protein recognition could eventually be carried out in a single step by directly grafting the double-stranded DNA already bound with the antibody of interest onto the SAM, and such advancement could significantly reduce both the procedural steps needed and relative handling time as well as the costs of analysis in the near future.

Having discussed the basic principles of miniaturization, and functionalization of micro- and nanodevices for biologic purposes, in the following sections a formal introduction to the state of the art of novel tools for in vitro and in vivo diagnostics will be offered.

2.4 In vitro Diagnostics

The current standard for immunoassays and emerging molecular diagnostics is to analyze one single analyte: specifically, one aliquot of a patient sample (serum, plasma, or tissue specimens) is processed and tested for one analyte at a time. Pressure for more accurate tests supported the extensive research to achieve state-of-the-art performance of relatively simple assays. As a reflection, ELISA is one of the most important biochemical techniques used mainly to detect the presence of antibodies or antigens in a sample based on antibody–antigen immunoreactions [47].

Due to its simplicity, low cost, easy reading, acceptability, and safety ELISA is widely used for detection of cancer protein markers, pathogens, and other proteins relative to various diseases with a detection limit from 0.1 ng to 1 μg mL^{-1} [48–50]. However, the knowledge of specific tumor protein markers is still limited; on the other hand the level of those markers is very low at the early stage of several cancers and therefore beyond the detection limit of the state-of-the art diagnostic techniques.

Moreover, the demand for parallel, multiplex analysis of protein biomarkers from ever smaller biospecimens, such as those obtained with computed tomography–guided and stereotaxis biopsies or after cell sorting of more abundant neoplastic specimens meant to identify tumor stem cells (i.e., CD133 glial cells), is an increasing trend for both fundamental biology and clinical diagnostics. The need for increasing the current detection sensitivity, while searching for key proteins in smaller sample volumes, is therefore among the main triggering factors that led to the development of methods for single-cell analyses. Attempts included miniaturization and mimicking of conventional proteomic protocols as well as exploration of novel ideas and techniques that enable new types of experiments, expanding the scientific field of "cells on a chip" in "single cells on a chip" [51].

2.4.1 *Trends in Single-Cell DNA Barcode Analysis*

The demand for parallel, multiplex analysis of protein biomarkers from even smaller biospecimens is an increasing trend for both fundamental biology and clinical diagnostics [52]. To date, the most highly multiplex protein assays rely on spatially encoded antibody microarrays; this approach capitalizes on the chemical robustness of DNA oligomer strands and on the reliable assembly of DNA-labeled structures via complementary hybridization [53]. Recently the miniaturization of conventional techniques let to the development of DNA barcode-type arrays at 10 times higher density than standard spotted microarrays, potentially enabling for high-throughput and low-cost measurements.

Generally speaking, the immunoassay region of the chip should ideally be a microscopic barcode customized for the detection of many proteins and/or for the quantization of a single or few proteins over a broad concentration range. Accordingly, using the DNA-encoded antibody library technique, Fan et al. developed and optimized an antibody array applied toward the measurement of a highly multiplex panel of proteins from small whole-blood specimens (range of μL obtained through a finger prick). The versatility of their barcode immunoassay is demonstrated by the ability to stratify cancer patients via multiple measurements of a dozen blood protein biomarkers for each patient [54].

This technique was further applied by coupling it with the immobilization of living cells, with an outlook for a multiplex assay of citoplasmic proteins. Shin and colleagues, for example, were able to detect simultaneously not only enzymes, such as phospho-extracellular signal–regulated kinase (ERK), but also receptors, such as the epidermal growth factor receptor (EGFR), both key nodes of the PI3K signaling pathway of several tumors, at concentrations of 10 to 1 ng/mL [53, 55].

The coupling of DNA barcode analysis with immobilization and characterization of few to single cells represented a remarkable boost for cancer proteomics; in fact among the advantages of single-cell analysis the main one is certainly the possibility to foster qualitative protein measurements but in a quantitative fashion [55]. One example relates to the interrogation of cross talk between signaling pathways within a cellular population as a paradigm to understand the overall tumor architecture: in their experiments Wang et al. assessed how cell–cell contacts and soluble factor signaling influence interaction among a glioma cell line [56]. In particular they interrogated the activity associated with PI3K signaling in a model of glioma cancer (U87 cells line) as a function of cell–cell separation: their results indicated that only a subpopulation of cells presented a constitutive activation of EGFR, while the majority did not. This finding confirmed that such approach allows not only for a thorough quantitative in vitro analysis of the proteome of few living cells but also for a simulation of their hypothetical behavior in vivo: In line with the hypothesis that the expression of EGFR in a subpopulation of cells represents a trigger for parenchymal invasion, its expression in the majority or the entirety of the tumor would not enhance tumorigenicity but instead would create a self-inhibiting state.

2.4.2 *Trends in Carbon Nanotubes and Their Use in Microfluidic Devices*

Carbon nanotubes (CNTs) since their discovery have been the focus of research on account of exceptional chemical, mechanical, and electrical properties that explain their potential use in electrochemical and optical biosensing. However, manipulation and processing of CNTs have been limited by their compatibility with other materials. Considerable efforts have therefore been devoted to the surface modification of CNTs to pave the way for many useful applications and to realize the potential applications in successful products, especially composite manufacturing. The chemical modification, dispersion, and solubilization of CNTs represent an emerging area in the research on nanotube-based materials [57].

Several research groups have reported successful and doable functionalization techniques for single-walled carbon nanotubes (SWCNTs) and multiwalled carbon nanotubes (MWCNTs). SWCNTs represent a highly effective means of transporting cargos of various sizes and types across the cell membrane and are gradually playing a bigger and more important role in the field of nanomedicine. CNTs are easily functionalized with DNA by either covalent or noncovalent means. DNA–CNT complexes have been exploited in biosensing applications for detection of ions, glucose, peroxide, etc. Morevoer, Ou et al. developed a novel nanoprobe based on SWCNTs and a fluorescent photosensitizer pyropheophorbide a (PPa) and used them for cancer cell imaging and therapy in vitro. In their assays phospholipids bearing polyethylene glycol–modified SWCNTs that can provide an interface for the conjugation of PPa were prepared by ultrasonication, and then the polyethylene glycol–modified SWCNTs were conjugated with PPa by using the covalent functionalization method to construct SWCNT-PEG-PPa nanoprobes. The functionalization and stability of SWCNTs was evidenced by ultraviolet-visible (UV-Vis) absorption spectra and fluorescence spectra, while imaging of cancer cells with SWCNT-PEG-PPa nanoprobes was confirmed using two cancer cell lines via laser scanning confocal microscope tests, and killing of cancer cells with SWCNT-PEG-PPa was demonstrated using cytotoxicity tests [58].

Moreover, CNTs are now entering the field of microfluidics, where there is a large potential to provide integrated, tailor-made nanotube columns by means of catalytic growth of the nanotubes inside the fluidic channels. To this regard, Mathur and colleagues recently explored the hot-embossing method for transferring vertically aligned CNTs into microfluidic channels, fabricated on PMMA. By this way patterned and unpatterned aligned CNTs were transferred on the PMMA microchannel with the aim to realize a microfluidic-based point-of-care device for blood filtration and detection of biomolecules such as red blood cells filtered using laminar flow through the microfluidic channels. Hopefully, these and other methodologies will soon provide new methods to increase the understanding of cellular pathways involved in many pathologies, so tremendous advances might be expected within a few years from now both in degenerative and neoplastic diseases [59, 60].

Finally, CNTs are advocated also for cell tracking by MRI, an emerging technique in which a contrast agent should label cells efficiently at potentially safe concentrations, have high relaxivity, and should prevent alterations to the cellular machinery. Avti et al. reported the cytotoxicity, cytocompatibility, and cell labeling efficiency in NIH/3T3 fibroblasts of novel SWCNTs synthesized using gadolinium nanoparticles as catalysts (Gd-SWCNTs) indicating that Gd-SWCNTs label cells efficiently at potentially safe concentrations and enhance MRI contrast without any structural damage to the cells. In their experiments cells incubated with the Gd-SWCNTs at concentrations between 1 and 10 μg/mL for 48 hours showed no change in viability or proliferation compared to untreated controls. Additionally, at these potentially safe concentrations, up to 48 hours, the cells showed no phosphatidyl serine externalization (preapoptotic condition), caspase-3 activity (point of no return for apoptosis), genetic damage, or changes in their division cycle. Localization of Gd-SWCNTs within the cells was confirmed by transmission electron microscopy (TEM) and Raman microscopy, showing 100% cell labeling efficiency, with a significant uptake of Gd-SWCNTs by the cells and a resulting fourfold increase in

MR signal intensities as compared to untreated cells [61]. These results easily give the flavor of the rapid evolution witnessed by nanodiagnostics, which is naturally inclined to translate discoveries from tools for in vitro detection to those for in vivo analyses.

2.5 In vivo Diagnostics

Theranostics (a word derived from the fusion of therapeutics and diagnostics) is a proposed process of diagnostic therapy for individual patients that relies on the concept of find, fight, and follow. To this regard, in vivo diagnostics is expected to provide data instantaneously from the patient and to allow for an ongoing disease development and therapy efficacy assessment [62]. Supported by remarkable public interest, the need for personalized medicine has widely created momentum to develop point-of-care diagnostics with higher sensitivity, specificity, and reliability than ever before.

Advancements in this research area have been fostered by the latest discoveries in imaging single molecules and in creating implantable devices. Therefore many miniaturization techniques previously described in this chapter represent the cornerstone upon which these diagnostic tools are based [63]. Within the next decade such devices are expected to result in a significant improvement in the living conditions of people who need constant medical monitoring.

On the other hand the fusion of molecular targeting with the existing diagnostic imaging armamentarium is pushing further the limits of sensitivity and specificity of conventional techniques such as computed tomography, MRI, and even nuclear medicine.

2.5.1 *Implantable Devices*

A variety of illnesses require continuous monitoring in order to have efficient illness intervention. Physicochemical changes in the body can signify the occurrence of an illness before it manifests. Even with the usage of sensors that allow diagnosis and prognosis of the illness, medical intervention still has its downfalls: late detection, for instance, can reduce the efficacy of therapeutics.

Furthermore, the conventional modes of treatment can cause side effects such as tissue damage (chemotherapy and rhabdomyolysis) and induce other forms of illness (hepatotoxicity). Biopsies are required to diagnose pathologies, but because of their invasiveness, they are useless for frequent patient monitoring, neither are they useful to adjust therapy according to the patient's clinical response to disease management; moreover most of the optical and electronic features of pathologic markers are not even observable in macroscopic samples. A completely new approach is therefore warranted. The ability to repeatedly sample the local environment for pathologic biomarkers, chemotherapeutic agents, and tumor metabolite concentrations could improve response monitoring, early detection of metastasis, and personalized therapy [64].

The state of the art of sensors capable of continuous measurement of analytes in biological media is represented by a wide range of different devices whose potential use in single cells, tissue slices, animal models, and humans has been demonstrated. To date sensors specific for glucose, lactate, glutamate, pyruvate, choline, and acetylcholine have been realized and tested in laboratory assays. Unfortunately the criteria for sensor performance are still not strictly described and uniformly accepted. As a result biosensor calibration still represents the limiting factor for translation in clinical practice. For instance, once the issues related to their biocompatibility were cleared, devices for continuous blood glucose monitoring were the first biosensors of this kind to reach the human trial stage. The development of these biosensors dates back to the early 1990s, but only in the past decade has their miniaturization led to promising subcutaneuous implantable devices for continuous glucose monitoring. Several reviews have recently described this long research journey and the way researchers have created the basis for a constant readout of blood glucose concentrations and from this success story the common route to implement the spectrum of other possible markers detected by those biosensors [65, 66].

With the aim to build superior health regimes and ensure good patient compliance these advances in biosensor design and sensing efficacy need now to be amalgamated with research in responsive drug delivery systems. The use of drug delivery systems enables the lowering of side effects, with subsequent improvement in patient compliance. Chronic illnesses require continuous monitoring and medical intervention for efficient treatment to be achieved. Therefore, designing a responsive system that will reciprocate to the physicochemical changes may offer superior therapeutic activity. In this respect, integration of biosensors and drug delivery is a proficient approach and requires designing an implantable system that has a closed-loop system. This offers regulation of the changes by means of releasing a therapeutic agent whenever illness biomarkers prevail. Detecting an illness before it manifests by means of biomarkers levels will optimize therapeutic dosing and improve the management not only of tumors but also of a handful of chronic illness [67].

2.5.2 *Improvements in Conventional Radiology*

Development in the area of quantum dots, functionalized nanoparticles, and nanopowders is driving a real revolution in the field of in vivo imaging [68]. Some of these nanoparticles exhibit vectorial character and can be used within the human body as markers in nuclear imaging techniques such as MRI. Nanosized contrast agents are significantly improving the range of application and resolution power of MRI. For example, Au_3CU hollow nanoclusters with an average diameter of 48.9 ± 19.1 nm and a shell thickness of 5.8 $\pm$ 1.8 nm have been developed. These bimetallic agents enhance the contrast of blood vessels and offer great potential for use as intravascular contrast agents in MRI angiography [69]. Colloidal magnetic nanoparticles represent another group of agents for the visualization of organs by MRI. They combine a small size with strong magnetism, have high biocompatibility, and can bind to the desired receptors through an active functional group. When coupled to cancer-targeting antibodies, nanocrystals showed huge advantages for monitoring in vivo human cancer cells implanted in live mice [70].

Another class of innovative nanoparticles is represented by superparamagnetic iron oxide nanoparticles: after linkage with a phosphorothiotate-modified oligodeoxynucleotide complementary to c-fos mRNA they showed to be effective in tracing neurodegenerative diseases by MRI techniques [71]. A well-established application of cells labeled with superparamagnetic iron oxide or ultrasmall superparamagnetic iron oxide in combination with MRI is the tracking of immune cells (mainly monocytes or macrophages) during the development of an inflammatory process. Recently this same technique was applied to the detection of apoptosis, angiogenesis, and tissue infiltration during the development of cancer; lastly it was also proposed for stem cell tracking [72–74].

Beside conventional computed tomography and MRI, positron emission tomography (PET) and other nuclear medicine techniques also are benefitting from the development of targeted in vivo nanoimaging methods. For example, while bioactive radiotracer molecules are required to visualize organs, by development of new nanotracers meant to label specific genes of interest, nanotechnology could address a previously unseen challenge.

2.6 Future Trends

The future of healthcare relies heavily on diagnostics: ideally, very early detection of unique molecular or protein patterns would allow for tailored treatments and management of many degenerative and neoplastic diseases [75]. Indeed, oncology represents the first and foremost field of application of these novel technologies, hopefully for the expected high impact, unfortunately for the lack of valid alternatives.

Recently, an expanded collection of novel markers has emerged from numerous avenues of research and holds potential to be deployed in clinical practice to improve classification and management of oncologic patients with an overall amelioration of their clinical course. Among them, newly discovered stem and progenitor cell markers are of particular interest because once clinically validated, they might aid in the differential diagnosis of these tumors as well as in the monitoring of their responses to therapy (see Fig. 2.6).

Intensive research efforts are recently attempting to uncover agents that may target subpopulations of these cells with high tumorigenic potential and increased resistance to current therapies. Along these lines, the cell surface marker, CD133, and other markers

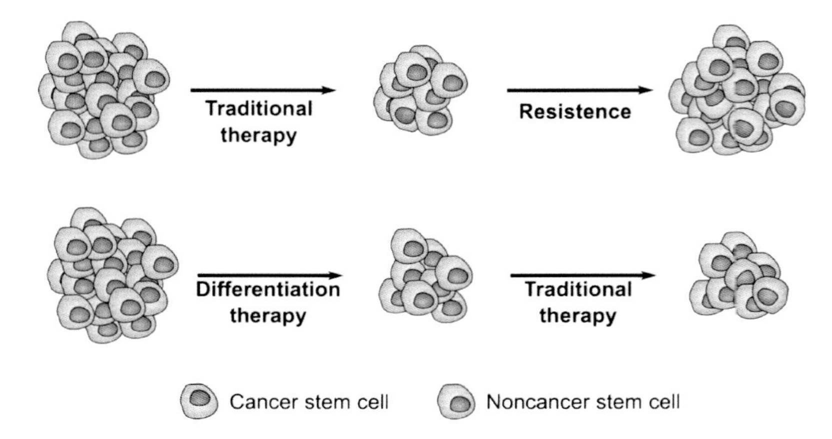

Figure 2.6 Role of stem cells in tumor recurrence.

of stem cells, such as Nestin, have been shown to negatively correlate with outcome parameters [76, 77]. To this regard Singh et al. showed that CD133+ brain tumor cells can self-renew and undergo lineage-specific differentiation [78].

However, not only primary tumors are under the magnification lens: most death in fact occurs from metastases to distant organs than from their primary localization. Such metastases arise whenever tumor cells escape the primary microenvironment and spread via the bloodstream. The study of such blood-borne cancer cells, also known as circulating tumor cells (CTCs), offers a unique window into the neoplastic diffusion process. Accordingly it is becoming abundantly clear that the biological and clinical value of CTCs exceeds their mere enumeration [79–81]. Emerging evidence indicates that CTCs are as much heterogeneous in nature as the primary tumor and may include a subset of cells that can successfully form metastases and/or cells capable of reseeding the primary tumor. For example, a recent study by Liu et al. demonstrated that human epidermal growth factor receptor 2 (HER2)-positive breast cancer patients with HER2-positive CTCs have longer progression-free survival after anti-HER2 therapies than HER2-positive patients with HER2-negative CTCs [82].

Unfortunately, most existing technologies do not allow for capture of live/viable cells that would enable the functional determination of such metastatic potential, aggressiveness, and chemotherapeutic sensitivity. Moreover, to date only epithelial markers have been extensively studies, and they are likewise ineffective in capturing cells that are losing epithelial properties and gaining mesenchymal characteristics [83]. Despite these current limitations, nanomedicine is not only adding a new layer of complexity to our understanding of cancer, but certainly it is also suggesting that we are on the right path to provide clinicians with highly useful tools for improving the diagnosis and classification of both primary and secondary tumors.

Acknowledgments

The present chapter is the result of a fruitful scientific collaboration with Stefania Corvaglia, Luca Ianeselli, and Maryse Nkoua (graduate students, NanoInnovation Lab at Elettra, Italy), Dr. Denis Scaini and Dr. Barbara Sanavio (graduate research assistant, NanoInnovation Lab at Elettra, Italy), and Dr. Anita Palma and Dr. Daniela Cesselli (Department of Medical and Biological Sciences, University of Udine, Italy). The deepest thanks go to Prof. Giacinto Scoles (Donner Professor of Science, Emeritus, Princeton University, USA), and Prof. Maurizio Fermeglia (Dean, University of Trieste, Italy).

References

1. Riehemann, K., Schneider, S. W., Luger, T. A., Godin, B., Ferrari, M., Fuchs, H. (2009) Nanomedicine-challenge and perspectives. *Angew Chem Int Ed Engl* **48**(5), 872–897.
2. Odreman, F., Vindigni, M., Gonzales, M. L., Niccolini, B., Candiano, G., Zanotti, B., Skrap, M., Pizzolitto, S., Stanta, G., Vindigni, A. (2005) Proteomic studies on low- and high-grade human brain astrocytomas. *J Proteome Res* **4**(3), 698–708.
3. Krutzik, P. O., Irish, J. M., Nolan, G. P., Perez, O. D. (2004) Analysis of protein phosphorylation and cellular signaling events by flow cytometry: techniques and clinical applications. *Clin Immunol* **110**(3), 206–221.

4. Leggett, G. J. (2006) Scanning near-field photolithography-surface photochemistry with nanoscale spatial resolution. *Chem Soc Rev* **35**(11), 1150–1161.
5. Truskett, V. N., Watts, M. P. (2006) Trends in imprint lithography for biological applications. *Trends Biotechnol* **24**(7), 312–317.
6. Kumar, A., Whitesides, G. M. (1993) Features of gold having micrometer to centimeter dimensions can be formed through a combination of stamping with an elastomeric stamp and an alkanethiol ink followed by chemical etching. *Appl Phys Lett* **63**, 2002–2004.
7. López, G. P., Biebuyck, H. A., Frisbie, C. D., Whitesides, G. M. (1993) Imaging of features on surfaces by condensation figures. *Science* **260**(5108), 647–649.
8. Bernard, M., Renault, J. P., Michel, B., Bosshard, H. R., Delamarche, E. (2000) Microcontact printing of proteins. *Adv Mater* **12**, 1067–1070.
9. Li, H. W., Muir, B. V. O., Fichet, G., Huck, W. T. S. (2003) Nanocontact printing: a route to sub-50-nm-scale chemical and biological patterning. *Langmuir* **19**, 1963–1965.
10. Mirkin, C. A., Letsinger, R. L., Mucic, R. C., Storhoff, J. J. (1996) A DNA-based method for rationally assembling nanoparticles into macroscopic materials. *Nature* **382**(6592), 607–609.
11. Piner, R. D., Zhu, J., Xu, F., Hong, S., Mirkin, C. A. (1999) "Dip-pen" nanolithography. *Science* **283**(5402), 661–663.
12. Chai, J., Wong, L. S., Giam, L., Mirkin, C. A. (2011) Single-molecule protein arrays enabled by scanning probe block copolymer lithography. *Proc Natl Acad Sci U S A* **108**(49), 19521–19525.
13. Liang, J., Castronovo, M., Scoles, G. (2012) DNA as invisible ink for AFM nanolithography. *J Am Chem Soc* **134**(1), 39–42.
14. Mirmomtaz, E., Castronovo, M., Grunwald, C., Bano, F., Scaini, D., Ensafi, A. A., Scoles, G., Casalis, L. (2008) Quantitative study of the effect of coverage on the hybridization efficiency of surface-bound DNA nanostructures. *Nano Lett* **8**(12), 4134–4139.
15. Castronovo, M., Scaini, D. (2011) The atomic force microscopy as a lithographic tool: nanografting of DNA nanostructures for biosensing applications. *Methods Mol Biol* **749**, 209–221.
16. Bosco, A., Bano, F., Parisse, P., Casalis, L., DeSimone, A., Micheletti, C. (2012) Hybridization in nanostructured DNA monolayers probed by AFM: theory versus experiment. *Nanoscale* **4**(5), 1734–1741.
17. Bano, F., Fruk, L., Sanavio, B., Glettenberg, M., Casalis, L., Niemeyer, C. M., Scoles, G. (2009) Toward multiprotein nanoarrays using nanografting and DNA directed immobilization of proteins. *Nano Lett* **9**, 2614–2618.

18. Sanavio, B., Scaini, D., Grunwald, C., Legname, G., Scoles, G., Casalis, L. (2010) Oriented immobilization of prion protein demonstrated via precise interfacial nanostructure measurements. *ACS Nano* **4**(11), 6607–6616.
19. Ling, M. M., Ricks, C., Lea, P. (2007) Multiplexing molecular diagnostics and immunoassays using emerging microarray technologies. *Expert Rev Mol Diagn* **7**(1), 87–98.
20. Rusling, J. F., Kumar, C. V., Gutkind, J. S., Patel, V. (2010) Measurement of biomarker proteins for point-of-care early detection and monitoring of cancer. *Analyst* **135**(10), 2496–2511.
21. Moncada, V., Srivastava, S. (2008) Biomarkers in oncology research and treatment: early detection research network: a collaborative approach. *Biomark Med* **2**(2), 181–195.
22. Wilson, R. (2013) Sensitivity and specificity: twin goals of proteomics assays. Can they be combined? *Expert Rev Proteomics* **10**(2), 135–149.
23. Czerkinsky, C. C., Nilsson, L. A., Nygren, H., Ouchterlony, O., Tarkowski, A. (1983) A solid-phase enzyme-linked immunospot (ELISPOT) assay for enumeration of specific antibody-secreting cells. *J Immunol Methods* **65**(1–2), 109–121.
24. Lees, J. E., Richards, P. G. (1999) Rapid, high-sensitivity imaging of radiolabeled gels with microchannel plate detectors. *Electrophoresis* **20**(10), 2139–2143.
25. Godovac-Zimmermann, J., Kleiner, O., Brown, L. R., Drukier, A. K. (2005) Perspectives in spicing up proteomics with splicing. *Proteomics* **5**(3), 699–709.
26. Kleiner, O., Price, D. A., Ossetrova, N., Osetrov, S., Volkovitsky, P., Drukier, A. K., Godovac-Zimmermann. J. (2005) Ultra-high sensitivity multi-photon detection imaging in proteomics analyses. *Proteomics* **5**(9), 2322–2330.
27. Arlett, J. L., Myers, E. B., Roukes, M. L. (2011) Comparative advantages of mechanical biosensors. *Nat Nanotechnol* **6**, 203–215.
28. Yang, Y. T., Callegari, C., Feng, X. L., Ekinci, K. L., Roukes, M. L. (2006) Zeptogram-scale nanomechanical mass sensing. *Nano Lett* **6**, 583–586.
29. Zougagh, M., Ríos, A. (2009) Micro-electromechanical sensors in the analytical field. *Analyst* **134**(7), 1274–1290.
30. von Muhlen, M. G., Brault, N. D., Knudsen, S. M., Jiang, S., Manalis, S. R. (2010) Label-free biomarker sensing in undiluted serum with suspended microchannel resonators. *Anal Chem* **82**, 1905–1910.
31. Nair, P. R., Alam, M. A. (2006) Performance limits of nanobiosensors. *Appl Phys Lett* **88**, 233120.

32. Sheehan, P. E., Whitman, L. J. (2005) Detection limits for nanoscale biosensors. *Nano Lett* **5**, 803–807.
33. Tamayo, J., Kosaka, P. M., Ruz, J. J., San Paulo, Á., Calleja, M. (2013) Biosensors based on nanomechanical systems. *Chem Soc Rev* **42**(3), 1287–1311.
34. Kusnezow, W., Syagailo, Y. V., Goychuk, I., Hoheisel, J. D., Wild, D. G. (2006) Antibody microarrays: the crucial impact of mass transport on assay kinetics and sensitivity. *Expert Rev Mol Diagn* **6**, 111–124.
35. Melli, M., Scoles, G., Lazzarino, M. (2011) Fast detection of biomolecules in diffusion-limited regime using micromechanical pillars. *ACS Nano* **5**(10), 7928–7935.
36. Picas, L., Milhiet, P. E., Hernández-Borrell, J. (2012) Atomic force microscopy: a versatile tool to probe the physical and chemical properties of supported membranes at the nanoscale. *Chem Phys Lipids* **165**(8), 845–860.
37. Liu, M., Amro, N. A., Liu, G. Y. (2008) Nanografting for surface physical chemistry. *Annu Rev Phys Chem* **59**, 367–386.
38. Liu, M., Liu, G. Y. (2005) Hybridization with nanostructures of single-stranded DNA. *Langmuir* **21**(5), 1972–1978.
39. Allen, S., Chen, X., Davies, J., Davies, M. C., Dawkes, A. C., Edwards, J. C., Roberts, C. J., Sefton, J., Tendler, S. J., Williams, P. M. (1997) Detection of antigen-antibody binding events with the atomic force microscope. *Biochemistry* **36**(24), 7457–7463.
40. Volkov, D., Strack, G., Halámek, J., Katz, E., Sokolov, I. (2010) Atomic force microscopy study of immunosensor surface to scale down the size of ELISA-type sensors. *Nanotechnology* **21**(14), 145503.
41. Schena, M., Shalon, D., Davis, R. W., Brown, P. O. (1995) Quantitative monitoring of gene expression patterns with a complementary DNA microarray. *Science* **270**(5235), 467–470.
42. Bernard, M., Renault, J. P., Michel, B., Bosshard, H. R., Delamarche, E. (2000) Microcontact printing of proteins. *Adv Mater* **12**, 1067–1070.
43. Niemeyer, C. M., Sano, T., Smith, C. L., Cantor, C. R. (1994) Oligonucleotide-directed self-assembly of proteins: semisynthetic DNA–streptavidin hybrid molecules as connectors for the generation of macroscopic arrays and the construction of supramolecular bioconjugates. *Nucleic Acids Res* **22**(25), 5530–5539.
44. Niemeyer, C. M., Boldt, L., Ceyhan, B., Blohm, D. (1999) DNA-Directed immobilization: efficient, reversible, and site-selective surface binding

of proteins by means of covalent DNA-streptavidin conjugates. *Ann Biochem* **268**(1), 54–63.

45. Niemeyer, C. M. (2002) The developments of semisynthetic DNA-protein conjugates. *Trends Biotechnol* **20**(9), 395–401.
46. Niemeyer, C. M. (2010) Semisynthetic DNA-protein conjugates for biosensing and nanofabrication. *Angew Chem Int Ed Engl* **49**(7), 1200–1216.
47. Jia, C. P., Zhong, X. Q., Hua, B., Liu, M. Y., Jing, F. X., Lou, X. H., Yao, S. H., Xiang, J. Q., Jin, Q. H., Zhao, J. L. (2009) Nano-ELISA for highly sensitive protein detection. *Biosens Bioelectron* **24**(9), 2836–2841.
48. Lequin, R. M. (2005) Enzyme immunoassay (EIA)/enzyme-linked immunosorbent assay (ELISA). *Clin Chem* **51**(12), 2415–2418.
49. Engvall, E., Perlmann, P. (1971) Enzyme-linked immunosorbent assay (ELISA). Quantitative assay of immunoglobulin G. *Immunochemistry* **8**(9), 871–874.
50. Koppelman, S. J., Lakemond, C. M., Vlooswijk, R., Hefle, S. L. (2004) Detection of soy proteins in processed foods: literature overview and new experimental work. *J AOAC Int* **87**(6), 1398–1407.
51. Lindstrom, S., Andersson-Svahn, H. (2011) Miniaturization of biological assays – Overview on microwell devices for single-cell analyses. *Biochim Biophys Acta* **1810**, 308–316.
52. Heath, J. R., Davis, M. E. (2008) Nanotechnology and cancer. *Annu Rev Med* **59**, 251–265.
53. Shin, Y. S., Ahmad, H., Shi, Q., Kim, H., Pascal, T. A., Fan, R., Goddard, W. A. 3rd, Heath, J. R. (2010) Chemistries for patterning robust DNA microbarcodes enable multiplex assays of cytoplasm proteins from single cancer cells. *ChemPhysChem* **11**(14), 3063–3069.
54. Fan, R., Vermesh, O., Srivastava, A., Yen, B. K., Qin, L., Ahmad, H., Kwong, G. A., Liu, C. C., Gould, J., Hood, L., Heath, J. R. (2008) Integrated barcode chips for rapid, multiplexed analysis of proteins in microliter quantities of blood. *Nat Biotechnol* **26**(12), 1373–1378.
55. Shi, Q., Qin, L., Wei, W., Geng, F., Fan, R., Shin, Y. S., Guo, D., Hood, L., Mischel, P. S., Heath, J. R. (2012) Single-cell proteomic chip for profiling intracellular signaling pathways in single tumor cells. *Proc Natl Acad Sci U S A* **109**(2), 419–424.
56. Wang, J., Tham, D., Wei, W., Shin, Y. S., Ma, C., Ahmad, H., Shi, Q., Yu, J., Levine, R. D., Heath, J. R. (2012) Quantitating cell-cell interaction functions with applications to glioblastoma multiforme cancer cells. *Nano Lett* **12**(12), 6101–6106.

57. Akbar, S., Taimoor, A. A. (2009) Functionalization of carbon nanotubes: manufacturing techniques and properties of customized nanocomponents for molecular-level technology. *Recent Pat Nanotechnol* **3**(2), 154–161.
58. Ou, Z., Wu, B. (2013) A novel nanoprobe based on single-walled carbon nanotubes/photosensitizer for cancer cell imaging and therapy. *J Nanosci Nanotechnol* **13**(2), 1212–1216.
59. Mathur, A., Roy, S. S., McLaughlin, J. A. (2010) Transferring vertically aligned carbon nanotubes onto a polymeric substrate using a hot embossing technique for microfluidic applications. *J R Soc Interface* **7**(48), 1129–1133.
60. Mogensen, K. B., Kutter, J. P. (2012) Carbon nanotube based stationary phases for microchip chromatography. *Lab Chip* **12**(11), 1951–1958.
61. Avti, P. K., Caparelli, E. D., Sitharaman, B. (2013) Cytotoxicity, cytocompatibility, cell-labeling efficiency, and in vitro cellular magnetic resonance imaging of gadolinium-catalyzed single-walled carbon nanotubes. *J Biomed Mater Res A.* DOI:10.1002/jbm.a.34643. [Epub ahead of print]
62. Heo, Y. J., Takeuchi, S. (2013) Towards smart tattoos: implantable biosensors for continuous glucose monitoring. *Adv Healthcare Mater* **2**(1), 43–56.
63. Wilson, G. S., Gifford, R. (2005) Biosensors for real-time in vivo measurements. *Biosens Bioelectron* **20**(12), 2388–2403.
64. Ishikawa, M., Schmidtke, D. W., Raskin, P., Quinn, C. A. (1998) Initial evaluation of a 290-microm diameter subcutaneous glucose sensor: glucose monitoring with a biocompatible, flexible-wire, enzyme-based amperometric microsensor in diabetic and nondiabetic humans. *J Diabetes Complications* **12**(6), 295–301.
65. Jeong, R. A., Hwang, J. Y., Joo, S., Chung, T. D., Park, S., Kang, S. K., Lee, W. Y., Kim, H. C. (2003) In vivo calibration of the subcutaneous amperometric glucose sensors using a non-enzyme electrode. *Biosens Bioelectron* **19**(4), 313–319.
66. Choleau, C., Klein, J. C., Reach, G., Aussedat, B., Demaria-Pesce, V., Wilson, G. S., Gifford, R., Ward, W. K. (2002) Calibration of a subcutaneous amperometric glucose sensor implanted for 7 days in diabetic patients. Part 2. Superiority of the one-point calibration method. *Biosens Bioelectron* **17**(8), 647–654.
67. Ngoepe, M., Choonara, Y. E., Tyagi, C., Tomar, L. K., du Toit, L. C., Kumar, P., Ndesendo, V. M., Pillay, V. (2013) Integration of biosensors and drug

delivery technologies for early detection and chronic management of illness. *Sensors (Basel)* **13**(6), 7680–7713.

68. Kinge, S., Crego-Calama, M., Reinhoudt, D. N. (2008) Self-assembling nanoparticles at surfaces and interfaces. *ChemPhysChem* **9**(1), 20–42.
69. Su, C. H., Sheu, H. S., Lin, C. Y., Huang, C. C., Lo, Y. W., Pu, Y. C., Weng, J. C., Shieh, D. B., Chen, J. H., Yeh, C. S. (2007) Nanoshell magnetic resonance imaging contrast agents. *J Am Chem Soc* **129**(7), 2139–2146.
70. Huh, Y. M., Jun, Y. W., Song, H. T., Kim, S., Choi, J. S., Lee, J. H., Yoon, S., Kim, K. S., Shin, J. S., Suh, J. S., Cheon, J. (2005) In vivo magnetic resonance detection of cancer by using multifunctional magnetic nanocrystals. *J Am Chem Soc* **127**(35), 12387–12391.
71. Liu, C. H., Huang, S., Cui, J., Kim, Y. R., Farrar, C. T., Moskowitz, M. A., Rosen, B. R., Liu, P. K. (2007) MR contrast probes that trace gene transcripts for cerebral ischemia in live animals. *FASEB J* **21**(11), 3004–3015.
72. Chang, H. H., Moura, J. M., Wu, Y. L., Ho, C. (2008) Automatic detection of regional heart rejection in USPIO-enhanced MRI. *IEEE Trans Med Imaging* **27**(8), 1095–1106.
73. Stoll, G., Bendszus, M. (2009) Imaging of inflammation in the peripheral and central nervous system by magnetic resonance imaging. *Neuroscience* **158**(3), 1151–1160.
74. Kim, J., Kim, D. I., Lee, S. K., Kim, D. J., Lee, J. E., Ahn, S. K. (2008) Imaging of the inflammatory response in reperfusion injury after transient cerebral ischemia in rats: correlation of superparamagnetic iron oxide-enhanced magnetic resonance imaging with histopathology. *Acta Radiol* **49**(5), 580–558.
75. Stegh, A. H. (2013) Toward personalized cancer nanomedicine: past, present, and future. *Integr Biol (Camb)* **5**(1), 48–65.
76. Rahman, M., Hoh, B., Kohler, N., Dunbar, E. M., Murad, G. J. (2012) The future of glioma treatment: stem cells, nanotechnology and personalized medicine. *Future Oncol* **8**(9), 1149–1156.
77. Mao, Y., Zhou, L., Zhu, W., Wang, X., Yang, G., Xie, L., Mao, X., Jin, K. (2007) Proliferative status of tumor stem cells may be correlated with malignancy grade of human astrocytomas. *Front Biosci* **12**, 2252–2259.
78. Singh, S. K., Clarke, I. D., Terasaki, M., Bonn, V. E., Hawkins, C., Squire, J., Dirks, P. B. (2003) Identification of a cancer stem cell in human brain tumors. *Cancer Res* **63**, 5821–5828.
79. Parkinson, D. R., Dracopoli, N., Petty, B. G., Compton, C., Cristofanilli, M., Deisseroth, A., Hayes, D. F., Kapke, G., Kumar, P., Lee, J. Sh., Liu, M. C., McCormack, R., Mikulski, S., Nagahara, L., Pantel, K., Pearson-White, S.,

Punnoose, E. A., Roadcap, L. T., Schade, A. E., Scher, H. I., Sigman, C. C., Kelloff, G. J. (2012) Considerations in the development of circulating tumor cell technology for clinical use. *J Transl Med* **10**, 138.

80. Cristofanilli, M., Braun, S. (2010) Circulating tumor cells revisited. *JAMA* **303**(11), 1092–1093.
81. Cristofanilli, M. (2009) The biological information obtainable from circulating tumor cells. *Breast* **18**(Suppl 3), S38–S40.
82. Liu, Y., Liu, Q., Wang, T., Bian, L., Zhang, S., Hu, H., Li, S., Hu, Z., Wu, S., Liu, B., Jiang, Z. (2013) Circulating tumor cells in HER2-positive metastatic breast cancer patients: a valuable prognostic and predictive biomarker. *BMC Cancer* **13**, 202.
83. Matthew, E. M., Gallant, J. N. (2013) Sizing up circulating tumor cells for personalized therapy. *Cell Cycle* **12**(15), 2.

Chapter 3

Surgery in the Realm of Nanometers

Mario Ganau,[a,b] Roberto Israel Foroni,[b,c] and Rossano Ambu[a]

[a]*Department of Surgical Science & Graduate School of Biomedical Engineering, University of Cagliari, Italy*
[b]*Minimally Invasive Robotic Surgery Lab, Department of Neurosurgery, University of Verona, Italy*
[c]*Brigham and Women Hospital, Boston, USA*
mario.ganau@singularityu.org

3.1 Introduction

Surgery involves the repair, resection, replacement, or improvement of body parts and functions in numerous ways, so often it is seen as a subspecialty of human engineering. Although the former statement is not completely true, there are certainly many areas in which surgical materials could be improved, but surgeons are generally unaware of materials available for use, while materials scientists do not know what surgeons require. As seen in the previous chapters, the combination of tissue engineering and nanomaterials has great potential for application to nearly every aspect of surgery. Noteworthy, tissue engineering will allow cells or artificial organs to be grown for specific uses, while nanotechnology will help to ensure maximal biocompatibility; biosensors will be combined

Commercializing Nanomedicine: Industrial Applications, Patents, and Ethics
Edited by Luca Escoffier, Mario Ganau, and Julielynn Wong

ISBN 978-981-4316-14-9 (Hardcover), 978-981-4613-14-9 (eBook)
www.panstanford.com

with improved electrodes and pacing devices to control impaired neurological functions. To explore all this area of interaction a Pubmed search was conducted for articles concerning the themes of "nanotechnology and surgery": it retrieved a total of 905 articles published since 1994. Accordingly, in this chapter we will review some of the areas where surgeons, engineers, and nanotechnologists have interacted in the past, and will discuss some of the most pressing problems that remain to be solved in the upcoming decades.

3.2 Potential Implications in the Field of Nanotechnology and Regenerative Medicine

Novel nanocomposite polymers are taking the stage in several areas of regenerative medicine and the speed by which new materials are proposed is increasing very fast. However, the development of synthetic tissues from polymeric materials, and their introduction in clinical practice, requires a thorough understanding of the basic mechanical and surface properties of each polymer, as the materials selection and processing techniques can affect the chemical, physical, mechanical, and cellular recognition properties of biomaterials. Furthermore, an obvious concern of any prospective medical device is the suitability of the constituent materials for the target tissue [1]. Specifically, evaluations of biocompatibility include identifying any acute toxins that have an immediate, detrimental effect on the host tissue, as well as longer-term responses of the host tissue resulting from the prolonged presence of the nanomaterial.

3.2.1 *Nanomaterials*

Polyhedral oligomeric silsesquioxane (POSS) with a distinctive nanocage structure consisting of an inner inorganic framework of silicon and oxygen atoms and an outer shell of organic functional groups is one of the most promising nanomaterials for medical applications [2]. Enhanced biocompatibility and physicochemical (material bulk and surface) properties have resulted in the development of a wide range of nanocomposite POSS copolymers for biomedical applications, such as the development of biomedical devices, tissue engineering scaffolds, drug delivery systems, dental applications, and biological sensors. The application of POSS nanocomposites in combination with other nanostructures has also been investigated, including silver nanoparticles and quantum dot nanocrystals. Chemical functionalization confers antimicrobial efficacy to POSS, and the use of polymer nanocomposites provides a biocompatible surface coating for quantum dot nanocrystals to enhance the efficacy of the materials for different biomedical and biotechnological applications. Interestingly, a family of POSS-containing nanocomposite materials can be engineered either as completely nonbiodegradable materials or as biodegradable materials with tunable degradation rates required for tissue engineering applications [3].

To give an example, due to its remarkable biocompatibility and in vivo biostability, a new nanocomposite polymeric material known as polyhedral oligomeric silsesquioxane-poly(carbonate-urea)urethane (POSS-PCU) has been proposed as an alternative to simple poly(carbonate-urea)urethane (PCU) for several surgical implants, including synthetic heart valve, lacrimal duct, bypass graft, and recently tracheal replacement. Actually, the reasons for this success are that the mechanical properties of POSS-PCU, including tensile strength, tear strength, and hardness, are much superior when compared to simple PCU. In fact, POSS-PCU (hardness 84+/−0.8 Shore A) demonstrates significantly higher tensile strength, 53.6+/−3.4 and 55.9+/−3.9 N mm^{-2} at 25°C and 37°C, respectively, than PCU, 33.8+/−2.1 and 28.8+/−3.4 N mm^{-2} at 25°C and 37°C, respectively; its tensile strength and elongation at break result significantly higher than PCU at both 25°C and 37°C. Moreover, POSS-PCU shows a relatively low Young's modulus (25.9+/−1.9 and 26.2+/−2.0 N mm^{-2}), which is significantly greater in comparison to PCU (9.1+/−0.9 and 8.4+/−0.5 N mm^{-2}) at 25°C and 37°C, respectively, with 100 μm thickness; and its surface appears significantly less hydrophilic than that of PCU, with relevant impact in terms of resistance to platelet adhesion [3].

Similar resistance to platelet adhesion was obtained by another group grafting a 2-methacryloyloxyethyl phosphorylcholine (MPC) onto a PCU surface via the Michael reaction. By this way, Gao et al. created a biomimetic structure characterized by a low water contact angle and high water uptake, resulting in improved surface hydrophilicity [4]. In addition, by imaging the grafted surface by atomic force microscopy (AFM) they showed that the grafted surface is rougher than the blank PCU surface, which plays a relevant role not only in terms of antithrombogenicity but also in terms of calcification resistance efficacy. These few examples, although very technical, give a flavor of the potential advantages in terms of long-term performances and durability of those nanomaterials and therefore of their potential application in any implanted biodevice, especially those with significant rate of blood contacting [5].

3.2.2 *Next Generation's Prostheses*

Tissue engineering is addressing many concerns regarding the overall performance of routine prosthetic implants, especially when compared to previous mechanical options, which often were inflexible and therefore very fragile. The shortcomings of conventional prosthesis, especially the metallic ones, are well known and include migration, subsidence, and stress shielding, as well as obscured postoperative radiologic assessment. Accordingly, the next generation of prosthesis, thanks to nanoderived smart materials and the subsequent possibility to enhance cell growth and integration, is rapidly finding its way in the clinical setting.

Mimicking the characteristics of a human tissue as complex as the bone matrix is of pivotal importance in prosthetic surgery. The bone acts as a nanocomposite with an organic (mainly collagen) and inorganic (nanocrystalline hydroxyapatite [HA]) components and a hierarchical structure ranging from the nano- to the macroscale; it provides mechanical support to our body, while transmitting physiochemical and mechanochemical cues. The clinical repair and reconstruction of acquired bone defects until recently has been conducted using autologous and allogenic tissues or alloplastic materials, with functional limitations. Therefore, the design and development of biomaterial scaffolds meant to replace the form and function of native bone while promoting regeneration without necrosis or scar formation is one of the most challenging area of research. Nanomaterials and nanocomposites are promising platforms to recapitulate the organization of natural extracellular matrix for the fabrication of functional bone tissues because their structure provides a closer approximation to native bone architecture [6].

Inspired by the hierarchical nanostructure of bone, the application of nanostructured materials for bone regeneration is gaining increasing interest, indeed. Unique properties of nanomaterials, such as increased wettability and surface area, lead to increased protein adsorption when compared to conventional biomaterials. Cell–scaffold interactions at the cell–material nanointerface may be mediated by integrin-triggered signaling pathways that affect cell behavior. For instance, supported by crystallographic and chemical studies showing that synthetic HA closely resembles the inorganic phase found in bones and teeth, nanostructured calcium phosphate ceramics have found application as bone fillers, cements, and coatings [7]. Similar nanostructured scaffolds have showed to provide structural support for autologous cells, while regulating their proliferation, differentiation, and migration, therefore supporting the formation of functional tissues. In fact, implant surfaces that better mimic the natural bone extracellular matrix can stimulate stem cell differentiation toward osteogenic lineages in the absence of specific chemical treatments. Both nanosized HA and β-tricalcium phosphate (β-TCP) have proven in vitro to increase the activity of alkaline phosphatase (ALP), an early marker of bone formation, and mRNA expression levels of osteoblast-related genes, such as the Runt-related transcription factor 2 (Runx-2) and bone sialoprotein (BSP), in total absence of osteogenic supplements [7, 8].

These nanomaterials have also been exploited in the design of cages with similar stiffness to human bone in order to reduce stress shielding of the inside graft, while stimulating adequate postoperative fusion. To this regard, polylactic acid (PLA) and its copolymer have a long history of safe clinical use; nevertheless PLA acidic degradation products can also cause asepsis inflammation, which could damage the microenvironment of bone formation. Initial laboratory tests conducted in the past decade showed that adding increasing percentages of β-TCP to a PLA polymer matrix stimulated the proliferation of human osteogenous cells and synthesis of the extracellular bone matrix in a dose-dependent manner. In vitro results indicated that in comparison to pure PLA, TCP-containing composite materials had faster degradation kinetics, caused less inflammatory reaction, and promoted contact osteogenesis. Finally, the composite material containing 60% β-TCP demonstrated a similar performance to pure TCP bone grafts in terms of osteogenesis, being compatible with the production of intraosseous implants for situations representing high levels of mechanical strain [9]. Recently, several in vivo studies have confirmed that the incorporation of β-TCP into PLA materials can both enhance its osteoconductivity and buffer acid products [10–12]. Additionally, nanosized β-TCP has shown improved mechanical properties and tunable degradability compared to microsized powders [12]. Therefore, these studies represent a success story of a potentially promising approach to design nanocomposite fusion devices combining the advantages of the two or more biodegradable materials and overcoming the disadvantages of each.

3.2.3 *Gene Therapy and Surgical Procedures*

Coronary stenting in percutaneous coronary intervention has revolutionized the field of cardiology because it represents a less invasive and costly procedure compared to coronary artery bypass graft. Nevertheless a major problem of coronary stenting is the risk of neointimal hyperplasia, which leads to restenosis and creates the basis for late stent thrombosis, two of the most serious and lethal complications of this procedure [13]. To overcome these complications drug-eluting stents were introduced into the market; currently five models are available in the U.S.: the CYPHER sirolimus-eluting stent from Cordis (approved by the Food and Drug Administration [FDA] in April 2003), the TAXUS Express and Liberté paclitaxel-eluting stents from Boston Scientific (approved by the FDA in March 2004 and October 2008, respectively), the ENDEAVOR zotarolimus-eluting stent from Medtronic (approved by the FDA in February 2008), and the XIENCE V everolimus-eluting stent from Abbott Vascular (approved by the FDA in July 2008). Clinical trials revealed that delivering drugs locally was effective in inhibiting neointimal hyperplasia, reducing the incidence of restenosis and providing a better safety profile as compared to radiation or systemic drug administration [14]. Following this trend, gene-eluting stents have recently been proposed as a novel method of circumventing rather than treating these problems. In fact utilizing nanotechnology, a sustained and localized delivery of genes can be obtained, thus mitigating all the problems of restenosis and late stent thrombosis because of an accelerated regenerative capacity of re-endothelialization [15]. To determine the potentiality of this technology clinical trial designs are currently underway, bearing in mind that as in most gene therapy protocols thorough long-term surveillance programs need to be cleared to ascertain its safety and efficacy.

3.2.4 *Preventing Postoperative Fibrosis and Excessive Cicatrization*

Surgical procedures might always be affected by excessive scar formation and postoperative fibrosis, as these inappropriate events occasionally occur after any surgical procedure, altering the cicatrization process, disturbing the postoperative course, and rendering reoperations more difficult and risky. Neurosurgical procedures (i.e., spinal interventions and craniotomies) are much affected by this adverse event: the literature describes this phenomenon as accompanying up to 20% of all neurosurgical procedures. The scar tissue that forms postoperatively adheres to the dura mater, penetrates into the spinal canal, and can cause narrowing symptoms, neurological deficits, and pain. The incidence and spread of this excessive scar or epidural fibrosis can be prevented through the modification of the surgical technique by incorporating endoscopic or microscopic access to minimize the operative field and the use of isolating substances (autogenous or heterogeneous) administered intraoperatively [16]. Aiming at reducing the immunological response and the local inflammatory process, laboratory tests have recently been conducted with the local use of membranes presenting a biodegradable nanofibrous net of poly(L-lactide-co-caprolactone) manufactured by an electro-spinning process. Experimental and control groups with rat models followed up for 30 days have confirmed that local cicatrization can be modified using nanomaterials in particular scar formation and

Table 3.1 Devices for nanosurgery

Nanodevice	Description
Cellular and subcellular surgery	
Nanoknives	Nanometer cutting-edge silicon nitride knife
Nanotweezers	Carbon nanotubes configured as pincers
Femtosecond lasers	Ultrashort laser pulse
Optical vortex trap	Vortex trap with optical tweezers
Nanoneedles	Nanoscale needle attached to atomic force microscope

epidural fibrosis can be limited and modified locally by preventing local inflammation processes [16].

3.3 Cellular and Subcellular Surgical Procedures

Delivering the surgical precision at the unprecedented cellular and subcellular length scale is the first and foremost mandate of nanosurgery; a future goal along this continuum of minimally invasive surgery is the development of nanodevices capable of performing surgery at a molecular or atomic level. In the next pages we will present some of the most reknowned technologies that are operational at the nanoscale level, such as nanoknives, nanotweezers, and femtosecond-laser systems. As deducible from Table 3.1 these nanodevices have the potential to revolutionize the practice of surgery in a profound and momentous way.

At present, nanoknives have been deployed and used effectively for microscale cellular surgery, allowing for precise targeted cutting. Among their greatest advantage is the possibility to constantly observe their tip in real time, allowing for used feedback, image capture, and even physiological recording (i.e., by electromyography techniques). Chang et al. examined the in vivo use of 10 to 100 μm long nanoknives with cutting edges of 20 nm in radius of curvature during experimental axonal reconstruction. Using those instruments they were able to make very small incisions (range of 50 to 100 μm long incisions) in nerve tissue in vivo and to repeat those incisions to progressively pare down the nerve as documented visually and by the accompanying incremental diminution of evoked motor responses recorded from target muscle [17]. Furthermore, these nanoknives showed to be safe in terms of induced neurotoxicity, as evidenced by the following robust growth of axons and neurons on this experimental material in vitro.

Subcellular surgery is much more complex. As described in the previous chapters, proteins and small-molecule metabolites constantly traffic among intracellular compartments, and it has become increasingly evident that biological specificity relies heavily on their spatial and temporal segregation and compartmentalization. Gaining information with regard to the spatial and temporal distribution and evolution of molecules within cells, therefore, is crucial to the construction of a quantitative model of cellular function. The ability to isolate selectively single subcellular compartments for chemical analysis or transplantation opens new venues for intervening in the cellular pathologies. Silicon nanotweezers (SNTs), for instance, are a well-known microsystem for molecular manipulation. They can be used to trap molecules while sensing their biomechanical and bioelectrical response in minute operations. SNTs can be mass-produced by highly parallel microsystem technology and their exploitation in single-cell nanosurgery has widespread applications in biology but so far has been limited by difficulty in maintaining the functionality of the transported subcellular organelles. This difficulty arises because of the propensity of optical tweezers to photodamage the trapped object. To minimize radiation damage to the trapped biological particle, laser wavelengths between ~800 nm and ~1100 nm are usually used because of the low absorption cross section of water and biological molecules in this spectral range. Nevertheless, at the high laser intensities required for trapping and translating subcellular organelles through the dense cellular microenvironment, photodamage via multiphoton processes is often inevitable. Jeffries et al. demonstrated that the use of polarization-shaped vortex traps alleviates this issue, because this strategy benefits from the highly nonlinear (fourth power for two-photon- and fifth power for three-photon-induced damage) dependence of photodamage on the sharpness of the intensity gradient [18]. The availability of such tools offers unprecedented control in active nanoscopic manipulation of subcellular structures and seems suitable also for in vitro manipulation of nanoparticles such as metal colloids and quantum dots.

Talking about single-cell laser nanosurgery, among the advantages are the following: (1) There is no physical contact with the cells, so they remain in a sterile environment; (2) there is high spatial selectivity so that single organelles can be extracted from specific areas of individual cells; (3) the method can be conducted in the cell's native media; and (4) in comparison to other techniques that target single cells, such as micromanipulators, laser nanosurgery has a comparatively high throughput.

3.3.1 *Design Improvements of Nanosurgical Instruments*

Beside few exceptions, such as SNTs and femtosecond-laser systems, much of the nanosurgical instrumentarium is still in the developing phase. Although those instruments serve the needs of proof-of-concept experiments, more refined equipment will need to be designed specifically to enable their efficient translation out of the laboratories in the surgical theater.

A fundamental question concerning the surgical use of microdevices that have the characteristic size of only several tens or hundreds of microns is whether such small instruments can be repeatedly used in an operatory field. For this nanoinstruments must be mechanically strong and ensure adequate performances. Commonly, the properties of nanomaterials provide significant advantages when compared to their conventional benchmarks: the ultimate strength of silicon nitride nanoknives, which range from 2 to 8 GPa, is actually stronger than bulk steel (on average 0.5 GPa). Moreover, the nanomaterials chosen for the fabrication of instruments designed for nanosurgery must not be subjected to plastic deformation, which could alter their performance after repeated use. As a result, although the manufacture of miniature-scaled surgical devices is based on well-developed and reasonably mature fabrication technology, the actual profile of such instrumentation in vivo and their potential use in cellular- and subcellular-scale surgical procedures have yet to be optimized.

Regarding the surgical manipulation of cells, one point concerns the development of miniaturized micromanipulators to economize on space utilization around the operative field; another area of critical need is the development of surgical microscopes with sufficient magnification to visualize down to unprecedented small scales (i.e., isolation of specific parts of neurons such as axons in peripheral nerve surgery for reconstruction of their functionality) and to provide optimal working distances between the optical elements and the tissue. For instance, Chang et al. suggest that the design of future surgical microscopes for cellular-scale neurosurgery should include on-board lighting and likely also incorporated fluorescence imaging, which provides more contrast and far better signal-to-noise ratios over bright-field imaging [17].

Another useful feature needed for a prompt diffusion of nanosurgery would be a mechanism to coordinate the spatial positioning and movement of the surgical field, the operating microdevices, and the field of view of the surgical microscope. Finally, ergonomic issues must be considered and incorporated into more refined systems in order to make them easy to use by the operator and therefore compatible with proper clinical use.

3.4 Nanorobotics

The use of swarms of nanorobots to perform seemingly miraculous tasks is a common trope in the annals of science fiction. Among the auspicated advantages of nanorobots these are the most important ones:

- As far as nanorobots are structured as nonbiological entities and do not generate any harmful activities, there shall be no side effects.
- Nanorobots could be useful in any kind of surgery both under general as well as regional anesthesia.
- Being highly specific and target oriented, they can reduce drug-related mortality and morbidity.
- As they bind the terminal receptors, there shall be no peaks and troughs in effects [19].

Although several of these remarkable features are still very much in the realm of fiction, the main disadvantage associated with nanorobots is the burden of initial high cost and complicated fabrication. Noteworthy, scientists have recently overcome many of the physical challenges associated with operating on the small scale and have generated the first generation of autonomous self-powered nanomotors and pumps.

3.4.1 *Miniaturized Propelling Systems*

Taking inspiration from biological motors, which have evolved over a million years to perform specific tasks with high efficiency, scientists are trying to create synthetic nanomachines mimicking the function of these amazing natural systems [20]. Living systems use biological motors to build life's essential molecules, such as DNA and proteins, as well as to transport those living blocks inside cells with both spatial and temporal precision. Some have even evolved sophisticated mechanisms to ensure quality control during nanomanufacturing processes, whether to correct errors in biosynthesis or to detect and permit the repair of damaged transport highways.

While artificial nanomotors pale compared to nature biomotors, recent advances indicate their great potential to perform diverse applications and demanding tasks [21]. Development of artificial nanomotors is still mainly at the microscale, helical swimmers are among the brightest examples. They possess several advantageous characteristics, such as high swimming velocity and precise motion control, indicating their potential for diverse applications, such as manipulation of small objects within liquid and flow control in lab-on-a-chip systems [22, 23]. Furthermore, when it comes to microsurgery it is essential to develop nanorobots that can be propelled wirelessly in fluidic environments with good control. To this regard, the latest prototypes of chiral colloidal propellers can be navigated in water with micrometer-level precision using homogeneous magnetic fields. Those propellers are made via nanostructured surfaces, can be produced in large numbers, carry chemicals, push loads, and act as local probes.

In a few experimental cases artificial nanomotors have reached velocities as large as 100 body lengths per second and relatively high powers to transport a "heavy" cargo within complex microchannel networks [24]. Despite this impressive progress, man-made nanomachines still lack the efficiency, functionality, and force of their biological counterparts and are limited to a very narrow range of environments. Improved understanding of the behavior of catalytic nanomotors will facilitate the design of highly efficient and powerful artificial nanomachines for complex operations in diverse realistic environments, making it easy to envision their practical application in the not-so-distant future [25].

3.4.2 *Artificial Cells*

Nanotechnology promises construction of artificial cells, enzymes, and genes. Thus it is spreading its wings to address the key problems in the field of replacement therapy for those disorders that are due to deficiency of enzymes, mutation of genes, or any repair in the synthesis of proteins. Currently nanodevices like respirocytes, microbivores, and probes encapsulated by biologically localized embedding have a greater application in treatment of anemia and infections. Here we provide a list of the first nanorobots proposed [26–30]:

- Vasculoids: A vasculoid is a single, complex, multisegmented nanotechnological medical robotic system capable of all transport functions of the blood, including circulation of respiratory gases, glucose, hormones, cytokines, waste products, and cellular components. This nanorobotic system could substitute the human vascular system. The vasculoid system conforms to the shape of existing blood vessels and serves as a complete replacement for natural blood.
- Respirocytes: These nanostructures transport oxygen in human body similar to erythrocytes. Originally proposed by Robert Frietas in 1996, they are less than a micrometer in diameter, possibly made of diamond, a biocompatible material. They also transport carbon dioxide. The proposed structure consists of three chambers: one would store oxygen, the other would store carbon dioxide, and the third would act as a buoyancy chamber, making the structure

floatable in blood. The structure would also have rotors to control the intake and exit of carbon dioxide and oxygen and for a controlled entry of glucose inside the structure to combine with oxygen and produce energy for the activity of respirocyte. Even a modification of these respirocytes is on a design proposal that can quickly remove certain poisonous substances from the body in poisoning patients (i.e., CO or nitrobenzene poisoning). As prolonged ventilator stay is a demon for anesthesiologists and intensive care specialists because of the associated risk of pneumonia and dependency, respirocytes have been proposed to speed up weaning from ventilators. Once they will be practically operational, they eventually would prove to be a pillar of strength to the intensivist.

- Clottocytes or artificial platelets would halt bleeding 100–1000 times faster than natural hemostasis. Besides being faster than natural platelets, the number required for the desired activity will also be less. Specially programmed motile clottocytes would even be able to detect internal bleeding and spontaneously seal the site, thus giving the hope that some emergency surgeries might be avoided. Also, this might help in bringing down the proportion of mortality in intensive care units that is yielded by silent uncontrolled bleeding.
- Microbivores (see Fig. 3.1) have been proposed to mimic white cells and perform phagocytosis of specific bacteria, viruses, or fungi. Today, physicians are facing the rapidly emerging problem of antibiotic resistance, many bacteria are resistant even to the highest class of antibiotics, and pan-drug resistance is becoming common on microbiology charts. The microbivore would be able to bind and target the pathogen with enzymes that would reduce the microorganism into basic amino acids, fats, and sugars harmless to the human body, thus offering a valid answer to rapidly evolving antibiotic resistance.

Until now, a popular dictum in science was that no machine is more efficient than the human body, but these upcoming inventions

Figure 3.1 Representation of a microbivore, an artificial white cell of a few nanometers, floating into the bloodstream.

are probably fated to change this assumption, as the advancement in their characterization is improving at a fast pace. If these prototypes will eventually prove to be effective not only in the laboratories but also at the bedsides, their performances will dramatically change the way we practice medicine [31].

3.5 Brain–Machine Interfaces

Conventionally, the practice of surgery has always been characterized by the removal of pathology, congenital or acquired. The emerging complements to the removal of pathology are procedures for the specific purpose of restoration of function. Indeed, advances in basic sciences along with a better understanding of anatomy and physiology are rapidly creating opportunities to intervene in

Table 3.2 Devices for biological interfaces

Nanodevice	Description
Microchips for biological interfaces	
Nanowires carbon nanotubes	Nanomechanical resonator
Nanoprobes	Nanocantilevers with biologic sensing
Nanobiosensors	Nanofibers for in vivo monitoring
Carbon transistors	Integrated logic circuit assembled on carbon nanotubes

disease processes in a reparative manner, thereby advancing toward this long-sought-after purpose. Accordingly, the next frontier for neurosurgery involves developing a greater understanding of the brain and furthering the surgeon's capabilities to directly affect brain circuitry and function. Soon, this has come in the form of implantable devices that can electronically and nondestructively influence the cortex and nuclei, with the purpose of restoring neuronal function and improving the quality of life [32, 33]. A few examples of such microchips for biological interfaces are reported in Table 3.2.

Recently, the arena of implantable devices has started approaching the issue of neurorestoration from a nanotechnological and biomedical engineering perspective. Devices such as deep brain stimulators, vagus nerve stimulators, and spinal cord stimulators are now becoming more commonplace in neurosurgery as we utilize our understanding of the nervous system to interpret neural activity and restore functions. One of the most exciting prospects in neurosurgery is the technologically driven field of the brain–machine interface, also known as the brain–computer interface, or neuroprosthetics. Once just in a prototype phase, brain–computer interfaces, which are machines that can decode the electrophysiological signals representing motor intent, are becoming concrete medical tools, due to the fast advances in molecular manufacturing and nanocomputation. In essence, these constructs can take some type of signal from the brain and convert that information into overt device control such that it reflects the intentions of the user's brain [34, 35].

Although the fields of nanotechnology and nanomedicine remain in their infancy at present the literature regarding nanoneurosurgery has rapidly expanded. The successful development of this technology will have far-reaching implications for patients suffering from a great number of diseases, including but not limited to spinal cord injury, paralysis, and stroke. Research in the area of microelectromechanical systems technologies is being principally conducted in the mesoscale, the realm between nanoscale and macroscale; although still larger than the nanoscale it seems, to date, easy to predict that these devices will certainly play an important role in the future of brain–computer interfaces. To this regard, the development of quantum computers and their potential to be thousands, if not millions, of times faster than current "classical" computers will significantly affect the neurosciences, especially the field of neurorehabilitation and neuromodulation. Quantum computers may advance our understanding of the neural code and, in turn, better develop and program implantable neural devices; when they will reach the point where we can actually implant such devices in patients, the possibilities of what can be done to interface and restore neural function will be limitless [36–38].

To resume, as applications of nanotechnology permeate all forms of scientific and medical research, new intriguing applications in the area of neurosciences will continue to emerge; therefore surgeons of the present and the future must take an active role in shaping the design and research of nanotechnologies to ensure maximal clinical relevance and patient benefit [39].

3.6 Conclusions

To conclude, surgery at the beginning of this millennium is characterized by an extremely rapid development, and there are no boundaries anymore among the different branches of basic sciences and medicine, nor between the living human and artificial tissues, or live surgery and virtual reality. Moreover, with the introduction of telesurgery there are no more surgical limits among countries and continents, as well as between earth and space. A new chapter in the history of medicine is ready to be written; as shown in the previous pages, nanotechnology and robotics are offering new unlimited possibilities for minimally invasive procedures. Although the future has already arrived, and is bringing a handful of new paradigms, the good old ethical requirements summarized in the Hippocrates Oath are to be fulfilled if we want to put the safety and freedom of our patients first [40].

Acknowledgments

This work has been written in the framework of the first author's PhD studies in Biomedical Engineering at the University of Cagliari, Italy. The deepest thanks go to Prof. Giacomo Cao, Prof. Gavino Faa, Prof. Alberto Maleci, and Prof. Franco Ennas.

References

1. C Grundfest-Broniatowski, S. (2013) What would surgeons like from materials scientists? *Wiley Interdiscip Rev Nanomed Nanobiotechnol* **5**(4), 299–319.
2. Ghanbari, H., Cousins, B. G., Seifalian, A. M. (2011) A nanocage for nanomedicine: polyhedral oligomeric silsesquioxane (POSS). *Macromol Rapid Commun* **32**(14), 1032–1046.
3. Kidane, A. G., Burriesci, G., Edirisinghe, M., Ghanbari, H., Bonhoeffer, P., Seifalian, A. M. (2009) A novel nanocomposite polymer for development of synthetic heart valve leaflets. *Acta Biomater* **5**(7), 2409–2417.
4. Gao, B., Feng, Y., Lu, J., Zhang, L., Zhao, M., Shi, C., Khan, M., Guo, J. (2013) Grafting of phosphorylcholine functional groups on polycarbonate urethane surface for resisting platelet adhesion. *Mater Sci Eng C Mater Biol Appl* **33**(5), 2871–2878.
5. Ghanbari, H., Kidane, A. G., Burriesci, G., Ramesh, B., Darbyshire, A., Seifalian, A. M. (2010) The anti-calcification potential of a silsesquioxane nanocomposite polymer under in vitro conditions: potential material for synthetic leaflet heart valve. *Acta Biomater* **6**(11), 4249–4260.
6. Kandziora, F., Pflugmacher, R., Scholz, M., Eindorf, T., Schnake, K. J., Haas, N. P. (2004) Bioabsorbable interbody cages in a sheep cervical spine fusion model. *Spine* **29**(17), 1845–1855.
7. Frost, A., Bagouri, E., Brown, M., Jasani, V. (2012) Osteolysis following resorbable poly-L-lactide-co-D, L-lactide PLIF cage use: a review of cases. *Eur Spine J* **21**(3), 449–454.

8. Huttunen, M., Ashammakhi, N., Törmälä, P., Kellomäki, M. (2006) Fibre reinforced bioresorbable composites for spinal surgery. *Acta Biomater* **2**(5), 575–587.
9. Aunoble, S., Clément, D., Frayssinet, P., Harmand, M. F., Le Huec, J. C. (2006) Biological performance of a new beta-TCP/PLLA composite material for applications in spine surgery: in vitro and in vivo studies. *J Biomed Mater Res A* **78**(2), 416–422.
10. Lin, K., Chang, J., Shen, R. (2009) The effect of powder properties on sintering, microstructure, mechanical strength and degradability of beta-tricalcium phosphate/calcium silicate composite bioceramics. *Biomed Mater* **4**(6), 65009.
11. Polini, A., Pisignano, D., Parodi, M., Quarto, R., Scaglione, S. (2011) Osteoinduction of human mesenchymal stem cells by bioactive composite scaffolds without supplemental osteogenic growth factors. *PLOS ONE* **6**(10), e26211.
12. McMahon, R. E., Wang, L., Skoracki, R., Mathur, A. B. (2013) Development of nanomaterials for bone repair and regeneration. *J Biomed Mater Res B Appl Biomater* **101**(2), 387–397.
13. Goh, D., Tan, A., Farhatnia, Y., Rajadas, J., Alavijeh, M. S., Seifalian, A. M. (2013) Nanotechnology-based gene-eluting stents. *Mol Pharm* **10**(4), 1279–1298.
14. Kollum, M., Bode, C. (2011) New developments in drug-eluting stents. *Herz* **36**(3), 177–188.
15. Doostzadeh, J., Clark, L. N., Bezenek, S., Pierson, W., Sood, P. R., Sudhir, K. (2010) Recent progress in percutaneous coronary intervention: evolution of the drug-eluting stents, focus on the XIENCE V drug-eluting stent. *Coron Artery Dis* **21**(1), 46–56.
16. Andrychowski, J., Frontczak-Baniewicz, M., Sulejczak, D., Kowalczyk, T., Chmielewski, T., Czernicki, Z., Kowalewski, T. A. (2013) Nanofiber nets in prevention of cicatrisation in spinal procedures. Experimental study. *Folia Neuropathol* **51**(2), 147–157.
17. Chang, W. C., Hawkes, E. A., Kliot, M., Sretavan, D. W. (2007) In vivo use of a nanoknife for axon microsurgery. *Neurosurgery* **61**(4), 683–691.
18. Jeffries, G. D., Edgar, J. S., Zhao, Y., Shelby, J. P., Fong, C., Chiu, D. T. (2007) Using polarization-shaped optical vortex traps for single-cell nanosurgery. *Nano Lett* **7**(2), 415–420.
19. Sengupta, S., Ibele, M. E., Sen, A. (2012) Fantastic voyage: designing self-powered nanorobots. *Angew Chem Int Ed Engl* **51**(34), 8434–8445.

20. Goel, A., Vogel, V. (2008) Harnessing biological motors to engineer systems for nanoscale transport and assembly. *Nat Nanotechnol* **3**(8), 465–475.
21. Wang, J. (2009) Can man-made nanomachines compete with nature biomotors? *ACS Nano* **3**(1), 4–9.
22. Zhang, L., Abbott, J. J., Dong, L., Peyer, K. E., Kratochvil, B. E., Zhang, H., Bergeles, C., Nelson, B. J. (2009) Characterizing the swimming properties of artificial bacterial flagella. *Nano Lett* **9**(10), 3663–3667.
23. Zhang, L., Peyer, K. E., Nelson, B. J. (2010) Artificial bacterial flagella for micromanipulation. *Lab Chip* **10**(17), 2203–2215.
24. Ghosh, A., Fischer, P. (2009) Controlled propulsion of artificial magnetic nanostructured propellers. *Nano Lett* **9**(6), 2243–2245.
25. Kostarelos, K. (2010) Nanorobots for medicine: how close are we? *Nanomedicine (London)* **5**(3), 341–342.
26. Agarwal, A. (2012) The future of anaesthesiology. *Indian J Anaesth* **56**(6), 524–528.
27. Freitas, R. A., Phoenix, C. J. (2002) Vasculoid: a personal nanomedical appliance to replace human blood. *J Evol Technol* **11**, 1–139.
28. Saha, M. (2009) Nanomedicine: promising tiny machine for the healthcare in future; a review. *Oman Med J* **24**, 242–247.
29. Freitas, R. A. (1998) Exploratory design in medical nanotechnology: a mechanical artificial red cell. *Artif Cells Blood Substit Immobil Biotechnol* **26**, 411–430.
30. Freitas, R. A. (2005) Microbivores: artificial mechanical phagocytes using digest and discharge protocol. *J Evol Technol* **14**, 55–106.
31. Freitas, R. A. (2002) The future of nanofabrication and molecular scale devices in nanomedicine. *Stud Health Technol Inform* **80**, 45–59.
32. Chen, Z., Appenzeller, J., Lin, Y. M., Sippel-Oakley, J., Rinzler, A. G., Tang, J., Wind, S. J., Solomon, P. M., Avouris, P. (2006) An integrated logic circuit assembled on a single carbon nanotube. *Science* **311**(5768), 1735.
33. Lee, B., Liu, C. Y., Apuzzo, M. L. (2013) A primer on brain-machine interfaces, concepts, and technology: a key element in the future of functional neurorestoration. *World Neurosurg* **79**(3–4), 457–471.
34. Elder, J. B., Liu, C. Y., Apuzzo, M. L. (2008) Neurosurgery in the realm of 10(-9), part 1: stardust and nanotechnology in neuroscience. *Neurosurgery* **62**(1), 1–20.

35. Elder, J. B., Liu, C. Y., Apuzzo, M. L. (2008) Neurosurgery in the realm of 10(-9), part 2: applications of nanotechnology to neurosurgery-present and future. *Neurosurgery* **62**(2), 269–284.
36. Leary, S. P., Liu, C. Y., Apuzzo, M. L. (2005) Toward the emergence of nanoneurosurgery: part I; progress in nanoscience, nanotechnology, and the comprehension of events in the mesoscale realm. *Neurosurgery* **57**(4), 606–634.
37. Leary, S. P., Liu, C. Y., Apuzzo, M. L. (2006) Toward the emergence of nanoneurosurgery: part II; nanomedicine: diagnostics and imaging at the nanoscale level. *Neurosurgery* **58**(5), 805–823.
38. Leary, S. P., Liu, C. Y., Apuzzo, M. L. (2006) Toward the emergence of nanoneurosurgery: part III; nanomedicine: targeted nanotherapy, nanosurgery, and progress toward the realization of nanoneurosurgery. *Neurosurgery* **58**(6), 1009–1026.
39. Lee, B., Liu, C. Y., Apuzzo, M. L. (2012) Quantum computing: a prime modality in neurosurgery's future. *World Neurosurg* **78**(5), 404–408.
40. Sándor, J., Máté, M., Irtó, I., Záborszky, A., Benedek, G., Sterlik, G., Regöly-Mérei, J. (2001) Surgery without boundaries. *Magy Seb* **54**(5), 303–306.

Chapter 4

Nanotherapeutics

Julielynn Wong
Center for Innovative Technologies and Public Health, Toronto, Canada
julielynnwongmd@gmail.com

4.1 Introduction

Nanotherapeutics apply the physical and chemical properties of nanomaterials (1–1000 nm in size) for the prevention and treatment of diseases [1]. Nanomaterials can be composed of metallic, organic, or semiconducting particles [2]. Nanotherapeutics are now being designed to shield drug agents against enzymatic degradation and immunogenicity, improve delivery of relatively insoluble drugs, facilitate the transportation of drugs across biological barriers, actively or passively target specific cells, deliver drugs to intracellular sites of action, apply diagnostic and therapeutic functions, and disrupt interactions between disease pathogen and host [1–5].

Nanotherapeutics have been applied for anesthesia, cancer, cardiovascular disorders, degenerative conditions, endocrine disorders, infectious diseases, and immune disorders [6]. The areas of greatest clinical impact of nanotherapeutics are in cancer and infectious diseases [7–11]. Nanovaccines are currently being

Commercializing Nanomedicine: Industrial Applications, Patents, and Ethics
Edited by Luca Escoffier, Mario Ganau, and Julielynn Wong

ISBN 978-981-4316-14-9 (Hardcover), 978-981-4613-14-9 (eBook)
www.panstanford.com

developed for hepatitis B, tuberculosis, leishmaniasis, diarrheal disease, and sexually transmitted diseases [12]. Nanovaccines are also being developed for applications outside of infectious diseases including type 1 diabetes, cancer, and tobacco cessation.

4.2 Biological, Physical, and Chemical Properties and Pharmacological Effects of Nanotherapeutics

Compared to their macrosized counterparts, nanotherapeutics are smaller in size, exhibit a high surface-area-to-volume ratio and can have adjustable optical, electromagnetic, and biological properties [13–14]. They can be designed to have different sizes, shapes, structures (hollow versus solid), chemical components, and surface chemical properties.

The small size of nanotherapeutics facilitates access to compartments of the human body that larger particles (>100 nm) cannot get into [5]. Nanoparticles between 10 and 100 nm in size can pass through capillaries without embolizing; avoid entrapment in the microstructures of the lung, liver and gallbladder; and escape microfiltration by the kidneys [1]. At this size range, nanoparticles are large enough to exhibit biological function, while having a prolonged circulation time relative to their macroscopic counterparts.

Because of their size, nanoparticles are optimized for drug inhalation therapy, a route of administration that offers the advantages of bypassing the "first pass effect" of orally ingested drugs, avoiding the risks of injected drug delivery, and circumventing the blood–brain barrier [15]. Nanoformulations can create stable aerosols and facilitate drug passage into pulmonary capillaries [16–20]. Targeted nanoparticles are being investigated for their ability to overcome cellular barriers to inhaled drug absorption through avoiding pulmonary clearance mechanisms or targeting of specific pulmonary cell types [21, 22].

As a result of their size, nanotherapeutics exhibit enhanced anticancer activity through a passive targeting mechanism known as the enhanced permeability and retention (EPR) effect [1, 2]. In normal tissue, the cells lining the interior of blood vessels are sealed and do not permit the passage of cells or particles. In contrast, the blood vessels of tumors are "leaky" which allows the entry of white blood cells since these vessels are in a heightened inflammatory state. This hyperpermeability permits the entry of nanoparticles ($<$200 nm in size) as well as the spread of metastatic cells. Tumors also have abnormal lymphatic vessels, which leads to ineffective drainage. This results in prolonged retention and accumulation of nanotherapeutics within the tumor.

As a particle decreases in size, a larger percentage of its atoms are found on its surface compared to its core [5]. This often leads to an increase in the particle's reactivity. A particle's total surface area also increases exponentially when particle size is reduced. So a reduction in drug particle size often renders it more reactive, increases its solubility, and correlates with better in vivo performance.

The high surface-area-to-volume ratio of nanomaterials permits their surfaces to be coated with a large number and multiple types of biorecognition molecules for selective targeting of diseased tissue, better biocompatibility, and reduced toxicity [2]. Targeting ligands, such as antibodies, RNA, DNA templates, or chemical affinities, can potentially improve drug uptake via cell receptor-mediated endocytosis [1, 3]. A single nanoparticle can be designed to contain multiple different targeting ligands. This capacity for multiple affinities allows nanotherapeutics to achieve high effective affinities, even if only low-affinity ligands are present.

Nanotherapeutics employ "Trojan horse" techniques (i.e., "stealth" packages) to cross biological barriers (i.e., drug resistance pumps or the blood–brain barrier) [1]. The blood–brain barrier is created through enzyme-based mechanisms and the tight junctions of endothelial cells. One nanoemulsion drug product is reportedly able to cross the blood–brain barrier [23]. Biodegradable polymeric nanoparticles coated with antibodies are being developed to be recognized by brain capillary receptors, pass through the blood–brain barrier, and deliver their drug contents to the central nervous system [24, 25].

Hollow or porous cores allow nanomaterials to encapsulate many drug particles in a single carrier unit [1]. The size, charge, surface properties, and ligand density and type will influence the pharmacokinetics of the nanovehicle. By enclosing the active agent within a protective nanoparticle shell, one can control the pharmacokinetics and biodistribution of the drug independent of the drug's chemistry. Relative to the payload capacity of other drug formulations, a nanoparticle can hold a large number of drug or gene molecules to deliver a highly concentrated dose to a cell. Nanoencapsulation strategies include polymers containing absorbed drugs, dendrimers or other nanobased structures chemically bound to drug agents, and liposomes or micelles containing nanodoses of toxic or insoluble agents [26–31].

Nanoformulations have been applied to aid in the delivery of potent but insoluble drug agents [32]. Many promising new drugs are hydrophobic, large molecular compounds that are derived from natural products. However, their insolubility leads to low effective plasma concentrations and impaired bioavailability. Nanoparticle delivery has proven effective for improving the delivery and performance of simvastatin (Zocor), a cholesterol-lowering statin oral medication that is derived from *Aspergillus terreus* [33].

Nanotherapeutics can be designed to be "smart," meaning they have the ability to respond to their environment. They can be designed to release drugs in response to local enzyme, chemical, thermal or photonic triggers.

The rate of drug release can also be adjusted by altering the composition of the nanobased drug carrier. Polymer-coated nanoparticles can block the adherence of opsonizing proteins and slow down drug half-life and clearance rates [2].

Nanotherapeutics could overcome drug resistance in microbes and cancer cells through three mechanisms [1]:

(i) Nanoparticles with specific surface coatings can bypass cell membrane transport-based drug resistance mechanisms and enter cells via endocytosis to deliver a highly concentrated drug dose within the cell.
(ii) Nanoparticles can encapsulate or be conjugated with drugs, so these drugs can avoid detection by recognition mechanisms used in drug resistance.
(iii) Nanoparticles can also carry compounds that block or are toxic to cell membrane pumps involved in drug resistance.

These features are being applied in different ways to an increasing number of drug delivery agents as well as combined platforms with both diagnostic and therapeutic functions [1]. Most approved nanotherapeutics ($n = 17$) employ passive targeting (i.e., size and geometry-dependent mode of action) [3, 6]. A 2013 review identified only one approved nanotherapeutic (Ontak, Seragen, Inc.) that employed active targeting, defined as "a mechanism of action beyond purely size-dependent biological or chemical interactions" [6]. Ontak is a protein nanoparticle that targets interleukin-2 to combat T-cell lymphoma. Scaling of functionalized nanoparticles has proven to be challenging due to the technical challenges and high costs of obtaining pure antibody preparations [1].

4.3 Nanoformulations

Colloids, emulsions, and gels are three nanoformulations that do not describe the features of the nanoparticles but refer to the physical state of the transport medium [1].

A colloid describes a mixture in which particles (colloid) are evenly distributed in a medium (dispersant) of a different substance. Colloidal nanoparticles are usually between 0.1 and 1000 nm in diameter. The suspended particles can range in size or weight or be uniform. Many nanoformulations are colloidal suspensions because this facilitates the delivery of insoluble agents [34, 35]. Amphotec (SEQUUS Pharmaceuticals Inc.) is Food and Drug Administration (FDA)-approved colloidal amphotericin B (115 nm), an antifungal agent administered subcutaneously to treat severe or life-threatening aspergillosis infections [6].

4.3.1 *Nanoemulsions*

Emulsions are colloidal mixtures containing liquid globules [1]. Emulsions contain two immiscible liquids phases and can be prepared through sonication. Emulsifiers (surfactants) can aid in the preparation of emulsions because they help create a boundary between the two liquid phases.

Neoral (Novartis Pharmaceuticals) is an FDA-approved cyclosporin nanoemulsion (10–150 nm), an immunosuppressant oral drug used to prevent organ rejection in heart, liver, or kidney transplant recipients [6]. Durezol (Siron Therapeutics, Inc.,) is an FDA-approved difuprednate ophthalmic nanoemulsion (110 nm) used to topically treat anterior uveitis, an inflammatory ocular condition. RESTASIS (Allergan) is an FDA-approved 0.05% cyclosporine nanoemulsion given topically to treat dry eyes. Liple (Mitsibushi Tanabe Pharma Corporation) is a palmitate alprostadil nanoemulsion (209 nm), which is approved for use in Japan for the intravenous treatment of peripheral vascular disease.

Forcetria (Novartis Pharmaceuticals) is an FDA-approved influenza A (H1N1) nanoemulsion (165 nm) vaccine adjuvant. Pandemrix (GlaxoSmithKline Biologicals) is an influenza A (H1N1) nanoemulsion (150–155 nm) vaccine adjuvant approved for use in the European Union.

Two billion people worldwide are infected with the cancer-causing hepatitis B virus, which leads to an estimated 600,000 deaths each year [36]. Needle-free hepatitis B vaccines with a nanoemulsion adjuvant, which contains emulsified detergent droplets 40 nm in size, are safe and effective in animal models [37]. This nasoemulsion is stable at room temperature for up to six months and does not exhibit the irritating side effects of standard adjuvants. The nasal delivery of this vaccine is pain free and avoids the risks associated with needles. As well, this vaccine series can be administered with a reduced dosing schedule [38]. This nanoemulsion strategy may someday prove effective for vaccines being developed against smallpox, influenza, anthrax, and human immunodeficiency virus (HIV) [12].

Nanoemulsion oral vaccines containing MAGE1-HSP70 and SEA complex proteins slowed tumor growth in mouse models [39]. Another nanoemulsion vaccine with a CpG ODN 1645 adjuvant reduced the size and occurrence of stomach cancer in mice [40].

Liposomes and micelles are specific types of emulsions in which the suspended globules are enveloped by an emulsification layer. Many nanotherapeutics in development employ nanoemulsion or liposomal formulations [6].

4.3.2 *Liposomes*

Liposomes are nanovesicles with a lipid bilayer encircling an aqueous core. They can contain a single (unilameral) or multiple layers (multilameral). They are synthesized through a number of methods, including sonication and electrolyte addition, extrusion, spray drying, or decompression [41–47].

Liposomes can solubilize drugs and those with phospholipid bilayers exhibit good biocompatibility and biodegradability [1]. Stealth liposomes have an inert PEG coating that prevents immune system recognition, which prolongs circulation time and promotes cell uptake. Liposomes can also be functionalized with targeting ligands. Targeted liposomes were initially developed nearly three decades ago [3]. The major drawbacks of liposomes are (1) their instability and (2) their ineffectiveness in drug-resistant conditions because they typically do not deliver drugs intracellularly. Despite this, liposome-based nanocarriers are commercially available worldwide [48].

Doxil (OrthoBiotech) is FDA-approved liposomal (87 nm) doxorubicin used to intravenously treat Kaposi's sarcoma, breast and ovarian cancer, and other solid tumors [6]. Evacet (The Liposome Company, Inc.,) is liposomal (150 nm) doxorubicin approved for intravenous use in the European Union. Myocet (Zeneus Pharma Sopherion Therapeutics) is a liposomal doxorubicin–citrate complex, which is a cardioprotective nanoformulation of doxorubicin [5]. It is given intravenously and is approved in Canada and the European Union. Lipo-Dox (Taiwan Liposome Company) is liposomal doxorubicin that is given intravenously to treat solid tumors. Lipo-Dox obtained foreign approval in 2001 [6].

DanuoXome (Gilead Sciences) is FDA-approved liposomal (45 nm) daunorubicin citrate, an intravenous chemotherapeutic used to treat solid tumors and HIV-related Kaposi's sarcoma. Diprivan (Zeneca Pharma) is FDA-approved liposomal (150–200 nm) propofol, a general anesthetic compound that is administered intravenously. AmBisome (Gilead Sciences) is FDA-approved unilameral liposomal amphotericin B (45–80 nm), an antifungal agent used to treat severe or life-threatening fungal infections. DepoCyt (SkyePharma Enzon) is FDA-approved sustained release liposomal cytarabine, which is administered intravenously for the treatment of lymphomatous meningitis [5].

4.3.3 *Micelles*

One can obtain more stable drug emulsions with drug-filled micelles that contain polymers in the micelle wall. These stable block copolymer micelles have been applied to many formulations [49–51], including chemotherapeutic agents and poorly insoluble drugs. One major advantage of block copolymer micelles is their success in helping agents evade immune system clearance. However, their drawbacks include micelle instability, low loading efficiency, poorly controlled drug release rates, and use of organic solvents during the production process.

SP1049C (Supratek Pharma Inc.,) is doxorubicin incorporated into block copolymer micelles and is FDA approved for the treatment of stomach cancer [52]. Estrasorb (Novavax) is an FDA-approved estradiol hemihydrate micelle, which is applied transdermally for the treatment of vasomotor symptoms in menopausal women [5].

Genexol-PM (Samyang) is being studied in advanced clinical trials (CT00912639) in over 20 patients with advanced solid tumors that were refractory to standard treatments [2]. Genexol-PM contains 20–50 nm micelles, which encapsulate paclitaxel, a cancer-fighting drug. The nanoparticle carrier altered the pharmacokinetics of paclitaxel, which permitted patients to receive a higher dose [53]. There appears to be no apparent side effects caused by the nanoparticle micelle carrier.

4.3.4 *Nanogels*

Nanogels are colloidal mixtures of a solid suspended in a fluid. They are composed of hydrogel nanoparticles that contain a network of hydrophilic cross-linked polymer chains, which are capable of absorbing large amounts of water to form a gel compound [1]. They can be prepared by precipitation polymerization, cooling or evaporation of an emulsion. With the removal of the fluid phase, solid gels are created with nanoscale voids, which can be utilized for drug delivery [54, 55]. Nanogels are biocompatible and biodegradable and can be designed to encapsulate a variety of bioactive compounds [56–58]. Nanogels can be also be smart and be designed to release their drugs in response to specific environmental stimuli.

4.3.5 *Dendrimers*

Dendrimers are branched nanostructures [2]. Their repeated branches provide multiple tunable sites for ligands (especially radioactive functional groups) for targeted and concentrated drug delivery [59–63]. Their uniform geometry gives rise to uniform dispersion rates [64–66]. Dendrimers have been tested as antibacterial drug carriers for tuberculosis and other mycobacteria [67–69].

Poly(amidoamine) (PANAM) is a commonly used dendrimer for nanobased drug delivery [70]. Targeting ligands and radioactive atoms can be attached to the PANAM dendrimer to deliver selective radiotherapy to tumors. One drawback of single-dendrimer compounds is that due to their size, they are eliminated by the liver and the reticuloendothelial system (RES) before they can accumulate in an effective concentration at their target sites.

Several clinical trials (ClinicalTrials.gov numbers NCT00370357, NCT00442910, and NCT00740584) are testing dendrimers that are designed to interfere with interactions between sexually transmitted disease pathogens and host. Dendrimers can inhibit the transmission of HIV in primates by binding to the virus and preventing it from entering the host cell [71]. The mode of inhibition varies according to the size and surface chemical functional groups of the dendrimer. Dendrimers with naphthalene disulfonate enters HIV cells and disrupts vital enzyme functions. Dendrimers with surface benzene dicarboxylate bind to the HIV viral capsid, which prevents the virus from entering the human host cell. Vivagel (Starpharma) is an investigational vaginal gel that contains a lysine-based dendrimer (5 nm) with surface naphthalene disulfonate, which delivers the microbicidal SPL7013 to prevent the transmission of HIV and herpes [72].

4.3.6 *Polymer Conjugates*

Polymer conjugates contain three structures: (1) a water-soluble polymer, (2) a linker, and (3) a bioactive agent (i.e., protein, low-molecular-mass drug, gene therapy) [1]. Polymers conjugates use chemical bonding to load drugs onto soluble polymer nanoparticles. Functional units can also be conjugated on the polymer in order to impart site-specific selectivity. Polymer conjugates can improve the delivery of gene, protein, and antibody-based drugs, which are often limited by a short plasma half-life, instability, or protein-induced immunogenicity.

Polymer selection is critical because the physical and chemical properties of the polymer often govern the distribution, metabolism and elimination of the conjugate. The most commonly used polymers in conjugates are poly(ethylene glycol) (PEG), *N*-(2-hydroxypropylmethacrylamide) (HMPA) copolymers, and poly(glutamic acid) (PGA).

The biodegradable polymer PEG is one of the most widely used polymers for protein conjugation, a process known as PEGylation. PEG conjugation improves protein stability and solubility, while reducing immunogenicity. PEGylation also prevents renal excretion and avoids receptor-mediated protein uptake (opsonization) by the RES, which prolongs plasma half-life and thereby enhances passive targeting effects. Through this mechanism, PEGylated drugs have a less frequent dosing schedule.

Polymer conjugates show promise for cancer treatment. They help solubilize hydrophobic chemotherapeutics (i.e., doxorubicin and paclitaxel), which facilitates systemic delivery. Polymer conjugates can be effective against cancer through the EPR effect and the lysosomotropic drug delivery pathway. Lysosomotropism is the action of drugs selectively placed in the intracellular partition between lysosomes and cytosol of targeted cells, which be beneficial for tumor destruction [73]. Improved anticancer activity and markedly reduced toxicity have been observed with polymer drug conjugates [74–76].

Oncaspar-IV (Rhone-Poulenc Rorer) is an FDA-approved agent containing polymeric nanoparticles bound to pegaspargase, which is given intravenously to treat acute lymphoblastic leukemia

[2]. Macugen (OSI Pharmaceuticals Pfizer) is an FDA-approved PEGylated anti-VEGF aptamer, which is administered intravitreally to treat neovascular age-related macular degeneration [5]. Adagen (Enzon) is an FDA-approved PEGylated adenosine deaminase, which is administered intravenously for enzyme replacement in severe combined immunodeficiency disease [6]. Pegasys (Nektar Hoffman-La Roche) is FDA-approved PEGinterferon alfa-2a, which is administered subcutaneously for treatment of chronic infection with hepatitis C. PEGIntron (Enzon Schering-Plough) is FDA-approved PEGinterferon alfa-2b, which is given subcutaneously for patients with chronic hepatitis C. Somavert (Nektar Pfizer) is FDA-approved PEGvesomant, which is administered subcutaneously for the treatment of acromegaly. Neulasta (Amgen) is FDA-approved PEG-G-CSF, which is administered subcutaneously for the treatment of febrile neutropenia. Renagel (Genzyme) is FDA-approved polymeric sevelamer HCL, which is taken orally for management of chronic kidney disease while on dialysis. CALAA-01 (Calando) is an investigational agent (NCT00689065) that contains polymeric nanoparticles (50–80 nm), which bind to transferrin receptors and deliver small interfering RNA that interferes with cancer gene expression [1].

The field of PEG–protein conjugates is expected to continue to rapidly expand [1]. Innovative forms of polymeric carriers are being developed for clinical testing. Polymer–protein conjugates are rapidly becoming an established class of antitumor agents.

4.3.7 *Gold Nanoparticles*

Gold nanoshells are quantum resonance particles that convert infrared light (800–1300 nm) into thermal energy [1]. Cells that are bound to the nanoshells are selectively destroyed by heat. Nanoshells consist of a dielectric core and a gold shell. The nanoparticle's specific optical resonant frequency is determined by the core–shell ratio. By altering the composition, size, and shell thickness, one can adjust the nanoshell's specific absorption profile.

Hyperthermia is a new form of minimally invasive cancer treatment in which thermal energy is applied for tumor ablation. This therapy requires controlled targeting of tumors and minimal heating of nearby healthy tissue. Gold nanoshells have been injected into recurrent head and neck tumors and then illuminated in the 700 to 800 nm wavelength range [77]. This leads to electron excitation and localized thermal ablation of tumors. This technique is currently under investigation in phase 1 clinical trials [2].

The photothermal approach has also been tested in mice using rod-shaped gold nanoparticles [78]. Gold nanorods possess properties that make them well suited for photothermal applications. Gold is biocompatible and can be easily functionalized to conjugate with specific biological targets. With an appropriate coating, they can have long plasma residence times. Gold nanorods have much greater absorption scattering properties compared to nanoshells. Unlike gold nanoshells, gold nanorods have minimal cytotoxicity because they do not interfere with DNA and RNA structures.

Gold nanoshells can be designed to come in stealth packages [79]. Researchers have induced monocytes to engulf coated gold nanoshells. These immune cells are released to invade a tumor. The gold nanoshells are then heated by optical stimulation. The monocytes act as "kamikaze" agents and destroy the local tumor cells.

Gold nanoparticles can be functionalized with a variety of ligands for use as sensors, drug carriers, and gene therapy [1]. Since folic acid receptors are overly expressed on cancer cells, gold nanoshells coated with folic acid with different PEG backbones have been used to target tumors for drug delivery and photothermal treatment [80].

Coated gold nanoparticles designed to target tumors in vivo may improve medication tolerance by reducing the entry of drugs into healthy cells. Aurimmune (CytImmune Sciences) is being tested in clinical trials (NCT00356980, NCT00436410) on patients with advanced or metastatic cancers that are no longer responsive to conventional therapies [2]. Aurimmune contains 27 nm sized gold nanoparticles that are coated with recombinant human tumor necrosis factor–alpha [TNF-alpha]) in order to minimize toxic side effects by reducing the accumulation of TNF-alpha in healthy tissue [2]. Patients who receive systemic infusions of Aurimmune are able to tolerate a twenty-fold higher dose of TNF-alpha [81, 82].

4.3.8 *Magnetic Nanoparticles*

Anticancer drugs could be bound to ferrofluids or nanomagnetic particles in order to concentrate the active agent to the tumor site by magnetic fields. The drug can be then detached from the magnetic carrier to act against the tumor [83–85]. Iron oxide nanoparticles (NanoTherm, Targeted Nano-Therapeutics) can be heated by an external magnetic field to deliver localized hyperthermic treatment for cancer [86, 87]. Magnetically heated iron oxide nanoparticles (MagForce NanoTherm) can provide hyperthermia therapy in combination with chemotherapy and radiation therapy for cancer treatment [88]. This combined approach leads to a lower dose of chemotherapeutics or radiation.

4.3.9 *Improved Cancer Therapies with Nanobased Drug Carriers*

The size, functionalization, and surface properties of nanoparticles can be fine-tuned to produce therapeutics targeted against specific tumor cells and with more selective pharmacological effects when compared to conventional agents [1]. Smaller-sized nanotherapeutics (10–100 nm) are more easily infused into and are slower to diffuse out of tumors (EPR effect), which have highly porous feeding vessels and poor lymphatic drainage [1, 2]. Antibody-coated nanoparticles designed to target tumors in vivo may improve medication tolerance by reducing the entry of drugs into healthy cells. Nanoparticles are being used to improve the delivery and performance of many highly potent anticancer agents derived from plant sources that are insoluble and toxic and cannot be effectively delivered across biological barriers. Much of the research and development in nanotherapeutics focuses on cancer treatment but most of this is still in the research phase [6]. Failure of nanotherapeutics in clinical testing remains a significant issue for research and development in the biotechnology and pharmaceutical industries [1].

Abraxane (Abraxis BioScience) is an FDA-approved nanoprotein-based systemic chemotherapy used to treat advanced lung and breast cancer [6]. Abraxane contains paclitaxel that is bound to albumin nanoparticles (130 nm), which minimizes the risk of an allergic reaction [89].

4.3.10 *Encapsulating Nanovaccines*

Encapsulating vaccine antigens into bioadhesive nanoparticles could potentially improve the efficacy of nasal vaccines [90]. Particulate adjuvants can also help to shield the antigen from degradation during intramuscular delivery and can also stimulate cell-mediated immunity for viral infections [12].

Biodegradable encapsulating nanoparticles, such as polylactide-co-glycolide acid (PLGA), have been developed as nonirritating adjuvants for polymer-based vaccines. The advantages of using PLGA nanoparticles in vaccines are biocompatibility, an excellent safety record, and a reduced dosing schedule because of sustained release of antigen [91].

An oral DNA vaccine for hepatitis B using encapsulated PLGA nanoparticles was shown to enhance immunity in animal models [92]. One dose of hepatitis B–encapsulated PGLA nanoparticles administered to mice could create immunity levels comparable to three doses of the standard hepatitis B vaccine [93].

Chlamydia trachomatis is the world's most common sexually transmitted disease, which leads millions worldwide to suffer from blindness, chronic pain, miscarriages, and sterility [94]. A PGLA85/15-encapsulated nanoparticle (~200 nm) vaccine against *C. trachomatis* (with 98% encapsulation efficiency) has been shown to generate an immune response in mice [95].

Enterotoxigenic *Escherichia coli* is a leading cause of diarrhea in the developing world, and an encapsulated vaccine has been developed and tested in mice [96]. This vaccine contains nanospheres (80–200 nm in size) that encapsulates two enterotoxins. Studies on mice showed that this vaccine generates a higher and longer immune response compared to a control.

Tuberculosis is the second leading cause of death by an infectious agent worldwide [97]. The Rv3619c nanovaccine encapsulates RD gene products and was developed for tuberculosis. This vaccine has been shown to generate an effective immune response and reduce disease burden in infected mice [98].

Leishmaniasis is a protozoan disease that affects an estimated 350 million people in 88 countries [99]. A single-dose nanovaccine has been developed against leishmaniasis, which contains chitosan nanoparticles that encapsulate a recombinant *Leishmania* superoxide dismutase. In mice, this vaccine can stimulate immune activity levels high effective for disease eradication [100].

4.3.11 *Nanobead Vaccines*

Nanobeads are defined as solid nanostructures whose surfaces are coated with antigens [12]. Their advantages include stability during storage, efficient uptake by antigen-presenting cells, and a lower dose of antigen to be used [101].

Nanobeads demonstrate a size-dependent effect on the immune system [102, 103]. Nanobeads 40 nm in size have been shown to promote higher T-cell responses, whereas nanobeads 50 nm in size are known to induce both B and T-cell-mediated immune responses in animal models.

Nanobead vaccines coated with type 1 diabetes-relevant peptide–major histocompatability complex molecules have been shown to decrease the autoimmune response against pancreatic beta cells and reverse the effects of type 1 diabetes on glucose control in mice [104].

4.3.12 *Micronanoprojection Vaccines*

Micronanoprojection vaccine delivery technology is painless and can more effectively deliver antigens to the highly immunogenic Langerhans cells that reside in the viable layer of the epidermis [12]. In 2008, Chen et al. developed a needle-free skin patch that contained microneedles with nanoscale tips coated with a DNA vaccine. These nanopatches coated with an influenza vaccine generated strong immune responses in mice [105].

4.3.13 *Combined Design Approaches for Nanovaccines*

A World Health Organization 2008 report estimated that 5.4 million deaths globally were caused by tobacco use. An effective nicotine nanovaccine could be of huge benefit to the millions of tobacco users who seek to quit and would generate billions of dollars in healthcare cost savings [106].

In 2011, phase 1 clinical trials began testing SEL-068, a self-assembling two-dose nanovaccine against nicotine in healthy smoker and nonsmoker adults (Selecta Biosciences 2011; ClinicialTrials.gov, 2012). SEL-068 is designed to prevent relapse during tobacco cessation [107]. SEL-068 utilizes both encapsulating nanoparticle and nanobead design to generate B- and T-cell-mediated immunity. No dose-limiting systematic toxicities were seen in preliminary safety studies in monkeys.

4.4 Commercial Advantages of Nanoformulated Agents

Nanoformulated drugs accounted for sales of $3 billion in 2006 according to *The Nanotech Report*, 5th edition, published by Lux Research. Nanotherapeutics offer new reformulation opportunities for active agents that were found to be insoluble or too toxic in standard oral or injectable forms [5]. Nanobased agents with superior pharmacokinetics and pharmacodynamics could be designed to have improved solubility, be delivered to specific intracellular targets, decrease immunogenicity, and reduce drug clearance rates. Therefore, failed agents could be nanoformulated to increase bioavailability, decrease toxicity, and improve compliance and clinical outcomes. In summary, nanotherapeutics have the potential to improve the efficacy of existing drug products, prolong the lifespan of proprietary drugs generate additional revenue streams, and reduce the time to market for active agents [1, 5]. Applying a nanoformulated approach could significantly impact the drug commercialization landscape.

4.5 Nanomedicine Publications, Patents, Product Development, and Companies

In the past two decades, the number of nanomedicine publications in the peer-reviewed literature has been steadily increasing, with accelerated growth starting around 2001 [108]. The number of nanomedicine patents has exploded since 2000. In 2005, over 200 companies had nanomedicine product sales totaling an estimated $6.8 billion.

A 2013 review identified 247 likely or confirmed nanomedicine applications or products [6]. One hundred sixty-nine clinical institutions (medical centers and universities) and companies are directly responsible for developing these products or applications. Fifty-four of these companies or institutions were developing more than one nanomedicine product or application. Over one-third of development in the nanomedicine sector was occurring at institutions and companies with only a single product or application. Most applications or products used nanomaterials at or below 200 nm in size. A steep drop-off in the number of nanomedicine products or applications was observed beyond phase II clinical trials.

4.6 Current Challenges and Priorities

Nanomedicine holds the potential to disrupt today's health care delivery model through impacting health economics and clinical outcomes. However, many challenges remain in bringing nanotherapeutics to patients. Seven priorities to support the commercialization of nanomedicines were identified through the March 2008 workshop convened by the FDA and the Alliance for Nanohealth in Houston, Texas [48]:

1. Determine the distribution of systemic nanoparticles in the body through all routes of administration.
2. Develop imaging techniques to visualize this distribution over time.
3. Understand how nanoparticles are transported across compartmental boundaries in the human body.
4. Validate new mathematical and computer models on the behavior of nanoparticles.
5. Create a "periodic table" of nanoparticles that predicts risk and benefit characteristics.

6. Establish benchmark standards for developing new classes of nanomedicines.
7. Create an analytic took kit for nanomedicine manufacturing, materials safety data sheets, and biodistribution profiles using standardized, validated techniques.

Nanomedicine faces an emerging number of translational, legal, ethical, safety, regulatory, and intellectual property issues [5, 48]. Experts agree that these challenges should be addressed through a broad collaborative private–public network of stakeholders, including industry, academia, professional organizations, patients and advocacy groups, regulatory bodies, policy makers, clinicians, technologists, toxicologists, molecular biologists, and medical scientists.

4.7 Nanotoxicity

While nearly 250 nanomedicine products are being used or tested in humans, it remains uncertain whether the increased production, handling and exposure of nanomaterials and by-products will lead to toxic effects in humans and the environment in the long term [6]. It cannot be assumed that nanomaterials will exhibit the identical toxicological profile of the same compounds at the macroscale [1].

To date, no conclusive evidence exists that demonstrates that nanomaterials cause toxicity in humans [2]. Many animal studies have shown that specific nanomaterials do not create toxic effects on liver and kidney function or histopathology analyses. More vital research is needed to understand the biodistribution, immunological response, processing, toxicity, and clearance of nanomaterials in the human body [109, 110].

Health and environmental toxicity concerns will impact the commercial success and public acceptance of nanotherapeutics. At present, stakeholders planning to commercialize, invest in, or regulate nanomedicines lack the resources to guide their decision making with respect to human and environmental health. Future research directed at understanding the effects of nanomaterials in vivo and in our environment will be critical for commercialization.

4.8 Future Trends

While most nanomedicine products approved before 2000 were nanotherapeutics, in the past decade, regulatory approval for nanotherapeutics has remained constant, while approval for nanobased devices has significantly increased [6].

While the most active area of research and development of nanotherapeutics is in drug delivery [5], nanobiologicals (sugar, proteins, nucleic acid, cell or tissue-based therapies) are predicted to expand in this sector [6]. Future work on nanovaccine development needs to demonstrate efficacy in human trials and address issues of quality control during the production process [12].

In vivo targeting will continue to be a major focus of development in nanomedicine [6]. Nearly 70 passive targeting products and 19 active targeting products are being clinically studied. Much work needs to be done to understand the role and importance of different factors for effective targeted drug delivery [111–113].

Combined nanomedicine treatments will become more common in the future. These could take the form of combining a nanotherapeutic with current therapies or using a single nanomedicine application for both diagnosis and treatment.

References

1. Tibbals, H. F. (2011) *Medical Nanotechnology and Nanomedicine*, CRC Press, Boca Raton, FL.
2. Kim, B. Y. S., Rutka, J. T., Chan, W. C. W. (2010) Nanomedicine. *N Engl J Med* **363**, 2434–2443.
3. Farokhzad, O. C., Langer, R. (2009) Impact of nanotechnology on drug delivery. *ACS Nano* **3**(1), 16–20.
4. Panyam, J., Labhasetwar, V. (2003) Biodegradable nanoparticles for drug and gene delivery to cells and tissue. *Adv Drug Delivery Rev* **55**, 329–347.

5. Bawa, R. (2010) Nanopharmaceuticals. *Eur J Nanomed,* **3**(1), 34–40.
6. Etheridge, M. L., Campbell, S. A., Erdman, A. G., Haynes, C. L., Wolf, S. M., McCullough, J. (2013) The big picture on nanomedicine: the state of investigational and approved nanomedicine products. *Nanomed: Nanotechnol, Biol, Med* **9**, 1–14.
7. Brown, D. M. (ed.) (2004) *Drug Delivery Systems in Cancer Therapy,* Humana Press, Totowa, NJ.
8. Peer, D., Karp, J., Hong, S., Farokhzad, O., Margalit, R., Langer, R. (2007) Nanocarriers as an emerging platform for cancer therapy. *Nat Nanotechnol* **2**, 751–60.
9. Kim, K. Y. (2007) Nanotechnology platforms and physiological challenges for cancer therapeutics. *Nanotechnol, Biol Med* **3**, 103–110.
10. Jain, K. K. (2008) Recent advances in nanooncology. *Technol Cancer Res Treat* **7**, 1–3.
11. Sahoo, S. K., Labhasetwar, V. (2003) Nanotech approaches to drug delivery and imaging. *Drug Discovery Today* **8**(24), 82–89.
12. Nandedkar, T. D. (2009) Nanovaccines: recent developments in vaccination. *J Biosci* **34**(6), 995–1003.
13. Xia, X, Xiong, Y. J., Lim, B., Skrabala, S. E. (2009) Shape-controlled synthesis of metal nanocrystals: simple chemistry meets complex physics? *Angew Chem Int Ed Engl* **48**, 60–103.
14. Peer, D., Karp, J. M., Hong, S., Farokhzad, O. C., Margalit, R., Langer, R. (2007) Nanocarriers as an amazing platform for cancer therapy. *Nat Nanotechnol* **2**, 751–760.
15. Illum, L. (2003) Nasal drug delivery: possibilities, problems, and solutions. *J Controlled Release* **87**, 187–198.
16. Geher, P., Muhlfeld, C., Rother-Rutishauser, B., Blank, F. (eds.) (2010) *Particle-Lung Interactions 2nd edition*, Informa Healthcare, New York, NY.
17. Shama, J. O.-H., Zhang, Y., Finlay, W. H., Roa, W. H., Löbenberg, R. (2004) Formulation and characterization of spray-dried powders containing nanoparticles for aerosol delivery to the lung. *Int J Pharm* **269**, 457–467.
18. Huang, Y. Y., Wang, C. H. (2006) Pulmonary delivery of insulin by liposomal carriers. *J Controlled Release* **113**, 9–14.
19. McMahon, G. T., Arky, R. A. (2007) Inhaled insulin for diabetes mellitus. *N Eng J Med* **356**, 497–502.
20. Sung, J. C., Pulliam, B. L., Edwards, D. A. (2007) Nanoparticles for drug delivery to the lungs. *Trends Biotechnol* **25**, 563–570.

21. Yang, W., Peters, J. I., Williams III, R. O. (2008) Inhaled nanoparticles: a current review. *Int J Pharm* **356**, 239–247.
22. Mistry, A., Stolnik, S., Illum, L. (2009) Nanoparticles for direct nose-to-brain delivery of drugs. *Int J Pharm* 379, 146–157.
23. Morgan, L. R., Struck, R. F., Waud, W. R., LeBlanc, B., Rodgers, A. H., Jursic, B. S. (2006) Carbonate and carbamate derivates of 4-demethylpenclomedine as novel anticancer agents. *Cancer Chemother Pharmacol* **64**(4), 829–835.
24. Jain, K. K. (2007) Nanobiotechnology-based drug delivery to the central nervous system. *Neuro-Degenerative Dis* **4**, 287–291.
25. Silva, G. A. (2007) Nanotechnology approaches for drug and small molecular delivery across the blood brain barrier. *Surg Neurol* **67**, 113–116.
26. Needham, D. (2003) Lipid membranes: biological inspiration for micro and nano encapsulation technologies, especially drug delivery. *Mater Res Soc Symp: Proc* **774**, 173–202.
27. Reis, C. P., Neufeld, R. J., Ribeiro, A. J., Viega, F. (2006) Nanoencapsulation I. Methods for preparation of drug-loaded polymeric nanoparticles. *Nanomed: Nanotechnol, Biol Med* **2**, 8–21.
28. Reis, C. P., Neufeld, R. J., Ribeiro, A. J., Viega, F. (2006) Nanoencapsulation II. Biomedical applications and current status of peptide and protein nanoparticulate delivery systems. *Nanomed: Nanotechnol, Biol Med* **2**, 53–65.
29. Torchilin, V. (2007) Micellar nanocarriers: pharmaceutical perspectives *Pharm Res* **24**, 1–16.
30. Huynh, N. T., Passirani, C., Saulnier, P., Benoit, J. P. (2009) Lipid nanocapsules: a new platform for nanomedicine. *Int J Pharm* **379**, 201–209.
31. Mishra, B., Patel, B. B., Tiwari, S. (2009) Colloidal nanocarriers: a review on formulation technology, types, and applications toward targeted drug delivery. *Nanomed: Nanotechnol, Biol, Med* **6**, 9–24.
32. Wang, J., Mongayt, D., Torchilin, V. P. (2005) Polymeric micelles for delivery of poorly soluble drugs: preparation and anticancer activity in vitro of paclitaxel incorporated into mixed micelles based on poly(ethylene glycol)-lipid conjugate and positively charged lipids. *J Drug Target* **13**, 73–80.
33. Park, J.-H., von Maltzhan, G., Ruoslahti, E., Bhatia, S. N., Sailor, M. J. (2008) Micellar hybrid nanoparticles for simultaneous magnetofluorescent imaging and drug delivery. *Angew Chem, Int Ed* **47**, 7284–7288.

34. Boyd, B. J. (2008) Past and future evolution in colloidal drug delivery systems. *Expert Opin Drug Delivery* **5**, 69–85.
35. Elaissari, A. (ed.) (2003) *Colloidal Biomolecules, Biomaterials, and Biomedical Applications*, CRC Press, Boca Raton, FL.
36. World Health Organization Media Centre. (2010) *Hepatitis B*. Retrieved from http://www.who.int/mediacentre/factsheets/fs204/en/
37. Makidon, P. E., Bielinska, A. V., Nigavekar, S. S., Janczak, K. W., Knowlton, J., Scott, A. J., et al. (2008) Pre-clinical evaluation of a novel nanoemulsion-based hepatitis B mucosal vaccine. *PLOS ONE* **3**, e2954. DOI:10.1371.
38. Pautler, M. Brenner, S. (2010) Nanomedicine: promises and challenges for the future of public health. *Int J Nanomed* **5** 803–809.
39. Ge, W., Li, Y., Li, Z. S., Zhang, S. H., Sun, Y. J., Hu, P. S., et al. (2009) The antitumour immune responses induced by nanoemulsion encapsulated MAGE1-HSP70/SEA complex protein vaccine following perioral administration route. *Cancer Immunol Immunother* **58**, 201–208.
40. Shi, R., Hong, L., Wu, D., Ning, X., Chen, Y., Lin, T., et al. (2005) Enhanced immune response to gastric cancer specific antigen peptide by coencapsulation with CpG oligodeoxynucleotides in nanoemulsion. *Cancer Biol Ther* **4**, 218–224
41. Basu, S. C., Basu, M. (eds.) (2002) *Liposome Methods and Protocols*, Humana Press, Totowa, NJ.
42. Agarwal, A., Lvov, Y., Sawant, R., Torchilin, V. (2008) Stable nanocolloids of poorly soluble drugs with high drug content prepared using the combination of sonication and layer-by-layer technology. *J Controlled Release* **128**, 255–260.
43. Leng, J., Egelhaaf, S. U., Cates, M. E. (2003) Kinetics of the micelle-to-vesicle transition: aqueous lecithin-bile salt mixtures. *Biophys J* **85**, 1624–1646.
44. Gregoriadis, G. (ed.) (2006) *Lipsome Technology: Volume I; Liposome Preparation and Related Techniques 3rd edition*, CRC Press, Boca Raton, FL.
45. Meure, L. A., Knott, R., Foster, N. R., Dehghani, F. (2009) The depressurization of an expanded solution into aqueous media for the bulk production of liposomes. *Langmuir* **25**, 336–337.
46. Mozafari, M. R., Mortazavi, S. M. (2005) *Nanoliposomes: From Fundamentals to Recent Developments*, Trafford Publishing, Oxford, UK.

47. Colas, J. C., Shi, W. L., Rao, V. S., Omri, A., Mozafari, M. R., Singh, H. (2007) Microscopical investigations of nisin-loaded nanoliposomes prepared by Mozafari method and their bacterial targeting. *Micron* **38**, 841–847.
48. Sanhai, W. R., Sakamoto, J. H., Canady, R., Ferrari, M. (2008) Seven challenges for nanomedicine. *Nat Nanotech* **3**, 242–244.
49. Khurana, S., Utreja, P., Tiwary, A. K., Jain, N. K., Jain, S. (2009) Nanostructured lipid carriers and their application in drug delivery. *Int J Biomed Eng Tech* **2**, 152–171.
50. Torchilin, V. P. (2004) Targeted polymeric micelles for delivery of poorly soluble drugs. *Cell Mol Life Sci* **61**, 2549–2559.
51. Ge, Z., Liu, S. (2009) Supramolecular self-assembly of nonlinear amphiphillic and double hydrophilic block copolymers in aqueous solutions. *Macromol Rapid Commun* **30**, 1523–1532.
52. Danson, S., Ferry, D., Alakhov, V., Margison, J., Kerr, D., Jowle, D., et al. (2004) Phase I dose escalation and pharmacokinetic study of pluronic polymer-bound doxorubicin (SP1049C) in patients with advanced cancer. *Br J Cancer* **90**, 2085–2091.
53. Kim, T. Y., Kim, D. W., Chung, J. Y., Shin, S. G., Kim, S. C., Heo, D. S., et al. (2004) Phase 1 and pharmacokinetic study of Genexol-PM, a cremophor-free polymeric micelle-formulated paclitaxel, in patients with advanced malignancies. *Clin Cancer Res* **10**, 3708–3716.
54. Chen, H., Chang, X., Du, D., Li, J., Xu, H., Yang, X. (2006) Microemulsion-based hydrogel formulation of ibuprofen for topical delivery. *Int J Pharm* **315**, 52–58.
55. Chen, H., Mou, D., Du, D., Chang, X., Zhu, D., Liu, J., et al. (2007) Hydrogel thickened microemulsion for topical administration of drug molecule at an extremely low concentration. *Int J Pharm* **341**, 78–84.
56. Raemdonck, K., Demeester, J., De Smedt, S. (2009) Advanced nanogel engineering for drug delivery. *Soft Matter* **5**, 707–715.
57. Oh, J. K., Lee, D. I., Park, J. M. (2009) Biopolymer-based microgels/nanogels for drug delivery applications. *Prog Polym Sci* **34**, 1261–1282.
58. Saunders, B. R., Laajam, N., Daly, E., Teow, S., Hu, X., Stepto, R. (2009) Microgels: from responsive polymer colloids to biomaterials. *Adv Colloid Interface Sci* **147–148**, 251–262.
59. Wang, H., Wang, S., Su, H., Chen, K-J., Armijo, A.-L., Lin, W.-Y., et al. (2009) A supramolecular approach for preparation of size-controlled nanoparticles. *Angew Chem, Int Ed* **48**, 4344–4348.

60. Das, A., Zhou, Y., Ivanov, A. A, Carter, R. L., Harden, T. K., Jacobson, K. A. (2009) Enhanced potency of nucleotide-dendrimer conjugates as agonists of the P2Y14 receptor: multivalent effect in G protein-coupled receptor recognition. *Bioconj Chem* **20**, 1650–1659.

61. Singh, P., Gupta, U., Asthana, A., Jain, N. K. (2008) Folate and folate-PEG-PAMAM dendrimers: synthesis, characterization, and targeted anticancer drug delivery potential in tumour bearing mice. *Bioconj Chem* **19**, 2239–2252.

62. Kono, K., Liu, M., Fréchet, J. M. J. (1999) Design of dendritic macromolecules containing folate or methotrexate residues. *Bioconj Chem* **10**, 1115–1121.

63. Woller, E. K., Cloninger, M. J. (2002) The lectin-binding properties of six generations of mannose-functionalized dendrimers. *Org Lett* **4**, 7–10.

64. Patri, A. K., Kukowska-Latollo, J. F., Baker Jr., J. R. (2005) Targeted drug delivery with dendrimers: comparison of the release kinetics of covalently conjugated drug and non-covalent drug inclusion complex. *Adv Drug Delivery Rev* **57**, 2203–2214.

65. Crampton, H. L., Simanek, E. E. (2007) Dendrimers as drug delivery vehicles: non-covalent interactions of bioactive compounds with dendrimers. *Polym Int* **56**, 489–496.

66. Das, J., Fréchet, J. M. J., Chakraborty, A. K. (2009) Self-assembly of dendronized polymers. *J Phys Chem B* **113**, 13768–13775.

67. Tomioka, H., Namba, K. (2006) Development of antituberculous drugs: current status and future prospects. *Kekkaku* **81**, 753–774.

68. Okada, M., Kobayashi, K. (2007) Recent progress in mycobacteriology *Kekkaku* **82**, 783–799.

69. Ma, M., Cheng, Y., Xu, Z., Xu, P., Qu, H., Fang, Y., et al. (2007) Evaluation of polyamidoamine (PAMAM) dendrimer as drug carriers of anti-bacterial drugs using sulfamethoxazole (SMZ) as a model drug. *Eur J Med Chem* **42**, 93–98.

70. Esfand, R., Tomali, D. A. (2001) Poly(amidoamine) (PAMAM) dendrimers: from biomimicry to drug delivery and biomedical applications. *Drug Discovery Today* **6**, 427–436.

71. Yu, J.-J., Nolting, B., Tan, Y. H., Xue, L., Gervay-Hague, J., Lu, G. (2005) Polyvalent interactions of HIV-gp120 protein and nanostructures of carbohydrate ligands. *NanoBioTechnology,* 201–210.

72. Owen, D. J. (2007) Dendrimers: new opportunities to treat and prevent human diseases. *Nanomed: Nanotechnol, Biol, Med* **3**, 338.

73. de Duve, C., de Barsy, T., Poole, B., Trouet, A., Tulkens, P., Van Hoof, F. (1974) Lysosomotropic agents. *Biochem Pharmacol* **23**, 2495–2531.

74. Duncan, R. (2006) Polymer conjugates as anticancer nanomedicines. *Nat Rev: Cancer* **6**, 688–701.

75. Cuchelkar, V., Kopecková, P., Kopecek, J. (2008) Synthesis and biological evaluation of disulfide-linked HPMA copolymer-mesochlorin e6 conjugates. *Macromol Biosci* **8**, 375–383.

76. Koppolu, B. P., Rahimi, M., Nattama, S. P., Wadajkar, A., Nguyen, K. (2009) Development of multiple-layer polymeric particles for targeted and controlled drug delivery. *Nanomed: Nanotechnol, Biol, Med* **6**, 355–361.

77. Hirsch, L. R., Stafford, R. J., Bankson, J. A., Sershen, S. R., Rivera, B., Price, R. E., et al. (2003) Nanoshell-mediated near-infrared thermal therapy of tumours under magnetic resonance guidance. *Proc Natl Acad Sci U S A* **100**, 13549–13554.

78. Huang, X., Neretina, S, and El-Sayed, M. A. (2009) Gold nanorods: from synthesis and properties to biological and biomedical applications. *Adv Mater* **21**, 2880–2910.

79. Choi, M. R., Stanton-Maxey, K. J., Stanley, J. K., Levin, C. S., Bardhan, R., et al. (2007) A cellular Trojan horse for delivery of therapeutic nanoparticles into tumours. *Nano Lett* **7**, 3759–3765.

80. Bhattacharya, R., Patra, C., Earl, A., Wang, S., Katarya, A., Lu, L., et al. (2007) Attaching folic acid on gold nanoparticles using noncovalent interaction via different polyethylene glycol backbones and targeting of cancer cells. *Nanomed: Nanotechnol, Biol, Med* **3**, 224–238.

81. Libutti, S. K., Paciotti, G. F., Myer, L., et al. (2007) Preliminary results of a phase 1 clinical trial of CYT-6091: a pegylated colloidal gold-TNF nanomedicine [abstract]. *J Clin Oncol* **25**(Suppl), 163s.

82. Visaria, R. K., Griffin, R. J., Williams, B. W., Ebbini, E. S., Paciotti, G. F., Song, C. W., et al. (2006) Enhancement of tumour thermal therapy using gold nanoparticle-assisted tumour necrosis factor-alpha delivery. *Mol Cancer Ther* **5**, 1014–1020.

83. Lübbe, A. S., Alexiou, C., Bergemann, C. (2001) Clinical applications of magnetic drug targeting. *J Surg Res* **95**, 200–206.

84. Häfeli, U., Schütt, W., Teller, J., Zborowski, M. (eds.) (1997) *Scientific and Clinical Applications of Magnetic Carriers*, Springer Verlag, Berlin, Germany.

85. Neuberger, T., Schopf, B., Hofmann, H., Hoffmann, M., von Rechenberg, B. (2005) Superparamagnetic nanoparticles for biomedical applica-

tions: possibilities and limitations of a new drug delivery system. *J Magn Magn Mater* **293**, 483–496.

86. Johannsen, M., Thiesen, B., Wust, P., Jordan, A. (2010) Magnetic nanoparticle hyperthermia for prostate cancer. *Int J Hypertherm*, 1–6.
87. Dennis, C. L., Jackson, A. J., Borchers, J. A., Hoopes, P. J., Strawbridge, R., Foreman, A. R., et al. (2009) Nearly complete regression of tumours via collective behaviour of magnetic nanoparticles in hyperthermia *Nanotechnology* **20**, 395103.
88. Thiesen, B., Jordan, A. (2008) Clinical applications of magnetic nanoparticles for hyperthermia. *Int J Hypertherm* **24**(6), 467–474.
89. Green, M. R., Manikhas, G. M., Orlov, S., Afanasyev, B., Makhson, A. M., Bhar, P., et al. (2006) Abraxane, a novel cremophor-free, albumin-bound particle form of paclitaxel for the treatment of advanced non-small-cell lung cancer. *Ann Oncol* **17**, 1263–1268.
90. Slutter, B., Hagenaars, N., Jiskoot, W. (2008) Rational design of nasal vaccines. *J Drug Target* **16**, 1–17.
91. Kersten, G., Hirschberg, H. (2004) Antigen delivery systems. *Expert Rev Vaccines* **3**, 453–462.
92. He, X., Wang, F., Jiang, L., Li, J., Liu, S. D., Xiao, Z. Y., et al. (2005) Induction of mucosal and systemic immune response by single dose oral immunization with biodegradable microparticles containing DNA encoding HBsAg. *J Gen Virol* **86**, 601–610.
93. Feng, L., Qi, X., Zhou, X., Maitani, Y., Wang, S., Jiang, Y. et al. (2006) Pharmaceutical and immunological evaluation of a single-dose hepatitis B vaccination using PLGA microspheres. *J Controlled Release* **112**, 35–42.
94. World Health Organization. (2012) *Initiative for Vaccine Research: Sexually Transmitted Diseases; Chlamydia trachomatis*. Retrieved from http://www.who.int/vaccine_research/diseases/soa_std/en/index1.html.
95. Taha, M. A., Singh, S. R., Dennis, V. A. (2012) Biodegradable PLGA85/15 nanoparticles as a delivery vehicle for Chlamydia trachomatis recombinant MOMP-187 peptide. *Nanotechnology* **23**, 325101. DOI:10.1088/0957-4484/23/32/325101.
96. Deng, G., Zeng, J., Jian M., Liu, W., Zhang, Z., Liu, X., et al. (2013) Nanoparticulated heat-stable (STa) and heat-labile B subunit (LTB) recombinant toxin improves vaccine protection against enterotoxigenic Escherichia coli challenge in mouse. *J Biosci Bioeng* **pii: S1389–1723**(12), 00377–5. DOI:10.1016/j.jbiosc.2012.09.009.

97. World Health Organization Media Center. (2013) *Tuberculosis*. Retrieved from http://www.who.int/mediacentre/factsheets/fs104/en.
98. Ansari, M. A., Zubair, S., Mahmood, A., Gupta, P., Khan, A. A., Gupta, U. D., et al. (2011) RD antigen based nanovaccine imparts long term protection by inducing memory response against experimental murine tuberculosis. *PLOS ONE* **6**, e22889.
99. http://www.who.int/leishmaniasis/en/
100. Danesh-Bahreini, M. A., Shokri, J., Samiei, A., Kamali-Sarvestani, E., Barzegar-Jalali, M., Mohammedi-Samani, S. (2011) Nanovaccine for leishmaniasis: preparation of chitosan nanoparticles containing Leishmania superoxide dismutase and evaluation of its immunogenicity in BALB/c mice. *Int J Nanomed* **6**, 835–842.
101. Gengoux, C., Leclerc, C. (1995) In vivo induction of CD4+ T cell responses by antigen covalently linked to synthetic microspheres does not require adjuvant. *Int Immunol* **7**, 45–53.
102. Fifis, T., Gamvrellis, A., Crimeen-Irwin, B., Pietersz, G. A., Li, J., Mottram, P. L., et al. (2004) Size-dependent immunogenicity: therapeutic and protective properties of nanovaccines against tumours. *J Immunol* **173**, 3148–3154.
103. Scheerlinck, J. P., Gloster, S., Gamvrellis, A., Mottram, P. L., Plebanski, M. (2006) Systemic immune responses in sheep, induced by a novel nanobead adjuvant. *Vaccine* **24**, 1124–1131.
104. Clemente-Casares, X., Tsai, S., Yang, Y., Santamaria, P. (2011) Peptide-MHC-based nanovaccines for the treatment of autoimmunity: a “one size fits all” approach. *J Mol Med (Berl)* **89**, 733–742.
105. Raphael, A. P., Prow, T. W., Crichton, M. L., Chen, X., Fernando, G. J., Kendall, M. A. (2010) Targeted, needle-free vaccinations in skin using multilayered, densely-packed dissolving microprojection arrays. *Small* **6**(16), 1785–1793.
106. DeBuono, B. A. (2006) *Milestones in Public Health: Accomplishments in Public Health over the Last 100 Years*, Pfizer Global Pharmaceuticals, New York, NY.
107. Kishimoto, T. (2012) Rational design of a fully synthetic nanoparticle-based vaccine [abstract], in. *Second Annual Conference on Vaccines and Vaccination*, Chicago, IL.
108. Wagner, V., Dullaart, A., Bock, A. K., Zweck, A. (2006) The emerging nanomedicine landscape. *Nat Biotechnol* **24**, 1211–1217.
109. Oberdörster, O., Stone, V., Donaldson, K. (2007) Toxicology of nanoparticles: a historical perspective. *Nanotoxicology* **1**, 2–25.

110. Vega-Villa, K. R., Takemoto, J. K., Yáñez, J. A., Remsberg, C. M., Forrest, M. L., Davies, N. M. (2008) Clinical toxicities of nanocarrier systems. *Adv Drug Delivery Rev* **60**, 229–238.

111. Dobrovolskaia, M. A. (2007) Immunological properties of engineered nanomaterials. *Nat Nanotechnol* **2**(8), 469.

112. Huang, X., Peng, X., Wang, Y., Wang, Y., Shin, D. M., El-Sayed, M. A., et al. (2010) A reexamination of active and passive tumour targeting by using rod-shaped gold nanocrystals and covalently conjugated peptide ligands. *ACS Nano* **4**(10), 5887–5896.

113. Walkey, C. D., Chan, W. C. W. (2011) Understanding and controlling the interaction of nanomaterials with proteins in a physiological environment. *Chem Soc Rev* **41**, 2780–2799.

SECTION II

Chapter 5

Ethics and Nanoethics

Mario Ganau,[a] Lara Prisco,[b] and Laura Ganau[c]
[a] *Kennedy Institute of Ethics, Georgetown University, Washington DC, USA*
[b] *University College Hospitals NSH Trust, London, UK*
[c] *School of Medicine, University of Cagliari, Italy*
mario.ganau@singularityu.org

5.1 Introduction

As seen in the previous section, nanotechnology holds the promise to redesign the realm of medicine as we know it nowadays, because it basically provides new dramatic tools to prevent diseases, promote health, and alleviate human suffering, which are surely among our strongest mandate. Having this huge impact, a potential for both great good or even great harm, the nanofuture certainly represents the latest stage toward which ethics is called to focus. Physicist Freeman Dyson once stated, "Progress of science is destined to bring enormous confusion and misery to mankind unless it is accompanied by progress in ethics" [1]. Ethics, in fact, has always provided a rational approach to moral dilemmas, and bioethics has expanded rapidly in recent decades to specifically address them in life sciences. Dilemmas in innovation technology are common, and

Commercializing Nanomedicine: Industrial Applications, Patents, and Ethics
Edited by Luca Escoffier, Mario Ganau, and Julielynn Wong

ISBN 978-981-4316-14-9 (Hardcover), 978-981-4613-14-9 (eBook)
www.panstanford.com

no big surprise if they are also gradually arising as part of the initial diffusion of nanotechnology-driven solutions to healthcare needs [2]. Aiming to better analyze and understand those dilemmas, while moving our scientific research at the maximum speed for the sake of its benefits, in this section of the book we will cover the themes of risk assessment, risk management, risk communication, and research and development regulations, and finally we will discuss the concerns associated with human enhancement.

5.2 Learning from Our Past

To start, let's raise a certain puzzle about our nanofuture. Defining in a nutshell the areas where nanomedicine will find its application is much complex because we have previously seen that they range from the medical applications of nanomaterials (for delivery of drugs, improvement of prosthesis, or enhancement of surgical tools), to nanoelectronic biosensors or neuroelectronic interfaces, to DNA repair through molecular nanotechnology [3, 4]. So what actually makes the era of nanomedicine so disruptive in terms of advancements when compared to the previous technological revolutions that the *ars medica* has witnessed in the past? Specifically, what pushes us to assume that the rise of this new nanotech era could represent a dramatic breakthrough in our history, even when compared to the recent hyperdebated biotech and genetic engineering waves? Is that their application in agrifood, which is particularly sensitive, seemed to provide no benefit to the consumers? Or is that the medical use of nanotechnologies will seem more acceptable to the people and will eventually make them see the related risks more favorably? We should recognize that those latter waves of biotech and genetech faced enormous obstacles in both governance (standards setting and regulatory issues) and social acceptance because they were perceived as "breaking species barrier" technologies [5–8]. To avoid repeating errors from the past, already in 2005 the National Science Foundation had tried to address these questions by organizing an international conference entitled "What Can Nano Learn from Bio"—essentially, the lesson for all future applications of nanomedicine is to ensure that public debate maintains a reasonable degree of context, which too often was lacking in the past [9].

Then, should nanomedicine be truly considered neutral, or are we underestimating its related externalities? Such a question, to date, has only speculative answers, and without a bioethical analysis most of us would struggle to propose rational, robust explanations. Despite the differences with biotech and genetech already mentioned, the principal feature of nanomedicine, which makes the case for discontinuity with previous technological waves, is the broader field of possible practical and theoretical applications. By entering the nanomedicine era we are accessing a funnel, from whom every initial step might be exponentially amplified down the road. Obviously, the point to undertake an ethical investigation at the very beginning of this era of nanomedicine is to address such awkward state of the art. In fact, along with the rise of any new emerging field of science and technology a careful reasoning helps in identifying the central ethical principles and precepts that must be prioritized in shaping the use of those discoveries and determining the right course of action to benefit from them [10].

5.3 Utopian Promises, Dystopian Fears

Despite its complex and often controversial nature, the starting point in our endeavor is certainly represented by a thorough risk assessment. We all are well aware that the fundamental questions behind each aspect of nanomedicine present ethical quandaries for the decision makers; thus an obvious question arises: might this analysis only come from those working in the field, or should also the general public being involved? This is not a rhetorical call: from novel diagnostic devices to performance-enhancing drugs, these dilemmas are impacting our goals, values, and aspirations as a society. It seems therefore better to get acquainted with the idea that all the concerned stakeholders, from patients to physicians, from big pharmas to their customers, are to be faced with those ethical questions, which if left unsolved, risk abruptly affecting our lives in the near future [11–13]. To prevent misuses of nanomedicine, and to guarantee that its whole set of advances would not eclipse the possible benefits, all the above-mentioned stakeholders are requested to be vigilant and proactively involved in a continuous constructive discussion meant to decide toward which directions we, as mankind, are willing to head.

Yet, part of the problem has been that many aspects, belonging to the sphere of nanoscale devices or drugs, have appeared so impenetrable, sometimes even shaded by a magic curtain, to the eyes of the public opinion. To this regard, we as scientists must admit that the actual purposes of nanomedicine have been rarely explained in a fashion really understandable to the general public, at least until recently. Therefore more efforts should be put into a new comprehensive approach, one in which the scientific community must be more focused on understanding all of the intricacies of the issues related to the long-term, large-scale diffusion of chip-on-a-lab essays, nanodrugs, etc., and then committed to share both their specific pros and cons with the final customers in order to obtain full awareness (if not proper informed consent) before their commercialization and use in clinical practice [14, 15].

Thus, an educational challenge is right upon us, and this book in part holds the ambition to properly start this path. Society has rapidly changed over the past decades, so thanks to the steps taken to encourage and foster scientific and humanistic studies in public schools, the general public is nowadays much better educated and therefore interested in every aspect of healthcare than it was years ago. Nevertheless building trust requires not only dedicated education but also strategic and intense outreach: these are the basis for an easier transition from an asymmetric to a more balanced transmission of relevant information between science or medical professionals and patients.

5.4 Development Regulations

Since fears about nanotechnology take many forms, mapping out nanorisks, especially those related to its possible toxicity, requires a strict law regulation of translational research. While some of the possible scenario sounds truly scary and scientifically unlikely, some others, however, are both quite realistic and troubling [2].

One particular hazard of coated nanoparticles is, for instance, that they can pass easily through most filtering methods currently in use, thus justifying the concerns for the environmental impact of those nanospheres, as pointed out by research works undertaken by insurance companies such as Swiss Reinsurance Company.

What could nanomedicine learn from previous genetic research? One answer is the widespread acceptation of endeavors such as the Human Genome Project, which made much progress also because it published the entire human DNA sequences on the Internet. This choice demonstrated that those sequences should not be kept private, nor tied to intellectual property, and ended up also proving to be effective in encouraging and nurturing a successful and proactive partnership between private and public institutions. To this regard, a continuing call for scientists to take greater part in considering the social equity, privacy, ethical, legal, safety, and environmental implications of their nanoscience is under our eyes. The time of engagement for upstream and transparent regulation of nanomedicine is upon us [9].

5.5 Evolving Our Nature

One final reason why we should care about our capacity to understand nanotechnological evolution and manage it properly also in terms of medical issues relates to a more philosophical concern—the possibility that the sophistication of nanomedicine, with specific regards for its neurotechnology implications, could one day affect the concept of free will by modifying some fundamental aspects of our nature, such as the way we percept sensations, feel emotions, store and retrieve our memories, or express our individuality and originality. As we will discuss whether the conditions for the boundaries of science to eventually transcend us as human beings would ever been met, several aspects of transhumanism with regard to the biological (current) limits of life will be the objects of this section [16–18]. Noteworthy, the lack of thorough, broad, and reliable scientific evidence that satisfactory experimental results are safely reproducible in a nonconstrained (out-of-lab) environment, to date, left us with many doubts on whether tech modifications of the human body will be practically feasible in the near future.

Transhumanists are attempting to spread their radical vision and could eventually impact nanotechnology's overall direction; but still we do not know whether their vision is more likely to cause problems than benefits for future generations. On the one hand, both the media and science fiction routinely remind us that several flaws in genetic determinism and posthumanism are possible and already in sight; on the other hand we would better consider that attempting to evolve ourselves without strict international consensus and appropriate rule of law is highly dangerous [18, 19].

As the lack of universality is destined to affect at least initially any new technology, which generally tends to be both very expensive and available only in certain countries, an unfair nanotech-induced evolution of our species unfortunately risks providing a biological basis for tech discrimination rather than futuristic social justice. Beside this terrific egalitarian problem, also the doubts concerning emotional and psychological drawbacks of transhumanism are far to be effectively answered. For these, declaring a moratorium (called for by rgw ETC-Erosion Technology and Concentration group yet in 2003) on a deregulated human enhancement must be our compelling mandate to avoid unforeseen and unintended consequences [20].

5.6 Conclusions

To conclude this introduction to nanoethics, let's remind once more that nanomedicine appears so radically innovative that it can really be compared to a train: in this analogy only those who are already on board may drive the train forward, whereas all the others still on the platform are debating pros and cons to decide whether to jump on board [21, 22]. In this light, we sincerely hope that this section will increase the readers' awareness, while helping them to take part in such a fundamental conversation meant to successfully overcome the strict dichotomy in thinking between a pure, old-fashioned, and useless science-driven-only or values-driven-only approach.

Acknowledgments

The present chapter has been written at the Kennedy Institute of Ethics, Georgetown University, Washington DC (U.S.). The deepest thanks go to Prof. Madison Powers, Dr. Laura Bishop, Dr. Martina Darragh, and Prof. James Giordano.

References

1. Dyson, F. (2006) *The Scientist as Rebel*, New York Review Books.
2. Collins, F. S. (2002) In *Cutting Edge Bioethics*, Kilner J. F., Hook C. C., Uustal D. B. (eds.), Eerdmans.
3. Resnik, D. B. (2012) Responsible conduct in nanomedicine research: environmental concerns beyond the common rule. *J Law Med Ethics* **40**(4), 848–855.
4. Kimmelman, J. (2012) Beyond human subjects: risk, ethics, and clinical development of nanomedicines. *J Law Med Ethics* **40**(4), 841–847.
5. Kuiken, T. (2010) Nanomedicine and ethics: is there anything new or unique? *Wiley Interdiscip Rev Nanomed Nanobiotechnol.* [Epub ahead of print] PubMed PMID: 20544800.
6. Sandhler, R., Kay, W. D. (2006) The GMO nanotech (dis)analogy? *Bull Sci Technol Soc* **26**, 57–62.
7. Sandler, R. (2009) Nanomedicine and nanomedical ethics. *Am J Bioeth* **10**, 16–17.
8. Khushf, G. Siegel, R. A. (2012) What is unique about nanomedicine? The significance of the mesoscale. *J Law Med Ethics* **40**(4), 780–794.
9. Kennetch, D., Thompson, P. B. (2008) *What Can Nanotechnology Learn from Biotechnology?* Elsevier.
10. Allhoff, F. (2009) The coming era of nanomedicine. *Am J Bioeth* **10**, 3–11.
11. Silva Costa, H., Sethe, S., Pêgo, A. P., Olsson, I. A. (2011) Scientists' perception of ethical issues in nanomedicine: a case study. *Nanomedicine (London)* **4**, 681–691.
12. King, N. M. (2012) Nanomedicine first-in-human research: challenges for informed consent. *J Law Med Ethics* **40**(4), 823–830.
13. Jones, R. (2008) When it pays to ask the public. *Nat Nanotechnol* **3**(10), 578–579.

14. Eaton, M. A. (2011) How do we develop nanopharmaceuticals under open innovation? *Nanomedicine* **4**, 371–375.
15. Marchant, G. E., Lindor, R. A. (2012) Prudent precaution in clinical trials of nanomedicines. *J Law Med Ethics* **40**(4), 831–840.
16. Lupton, M. (2007) Nanotechnology-salvation or damnation for humans? *Med Law* **2**, 349–362.
17. Thompson, R. E. (2007) Nanotechnology: science fiction? Or next challenge for the ethics committee? *Physician Exec* **33**(3), 64–66.
18. Nordmann, A. (2007) Knots and strands: an argument for productive disillusionment. *J Med Philos* **32**(3), 217–236.
19. Best, R., Khushf, G. (2006) The social conditions for nanomedicine: disruption, systems, and lock-in. *J Law Med Ethics* **34**(4), 733–740.
20. http://www.etcgroup.org/
21. White, G. B. (2009) Missing the boat on nanoethics. *Am J Bioeth* **10**, 18–19.
22. Khushf, G. (2007) Upstream ethics in nanomedicine: a call for research. *Nanomedicine (London)* **4**, 511–521.

Chapter 6

Nanomedicine Policy and Regulation Schemes

Sarah Rouse Janosik
Amgen Inc., Thousand Oaks, USA
sjanosik@amgen.com

6.1 Overview

The rapid expansion of the nanotechnology and nanomedicine markets presents complex policy and regulatory issues that traditional schemes are ill-equipped to sufficiently address. In addition, existing statutes and regulations leave governments and regulatory agencies unable to effectively respond to the challenges posed by nanomaterials used for medical applications [1]. Regulatory frontiers in the field of nanomedicine include areas such as product safety, privacy and civil liberties, occupational health and safety (OH&S), intellectual property (IP), international law, and environmental law, among others [2].

Uncertainty as to the potential health risks stemming from nanotechnology-based products has led some observers to suggest a deliberate slowing of nanotechnology research and development

Commercializing Nanomedicine: Industrial Applications, Patents, and Ethics
Edited by Luca Escoffier, Mario Ganau, and Julielynn Wong

ISBN 978-981-4316-14-9 (Hardcover), 978-981-4613-14-9 (eBook)
www.panstanford.com

until the development and implementation of nanotechnology-specific policies and regulations [3].

The span of nanotechnology across numerous disciplines and global markets demands that a range of international organizations and legal instruments contribute to the compilation of a cumulative and universal nanotechnology and nanomedical regulation mechanism [2]. Defining the universe of products subject to nanotechnology and nanomedical regulation alone will require enormous efforts. Consideration of a more functional definition of nanotechnology and nanomedicine may also be necessary [1]. The resulting standards and regulatory mechanisms will reduce ambiguity in nanomedicine and enable public acceptance and legitimation of the field.

6.2 Standardization of Nanotechnology Terminology and Characterization Methodologies

The development of nanotechnology and nanomedical products has progressed rapidly and hundreds of nanomedical-based products are now commercially available [4]. Despite the revolutionary potential of nanomedicine and the preclinical, clinical, and commercial successes demonstrated by various nanomedical technologies, the terminology and methodology used to describe and characterize underlying nanoenabled therapeutics remains at best inconsistent and at worst obstructive.

The lack of consistency in defining and characterizing nanomaterials and nanodevices for medical applications is primarily attributable to the fact that nanomaterials have no common properties other than a size of about 1–100 nm. The lack of colloidal stability in nanoparticle suspensions, agglomeration, polydispersity in nanoparticle size and shape, swelling, and leakage of encapsulated materials may also contribute to lack of consistent characterization methodology [5]. Other issues include difficulty of synthesis and processing techniques, inadequate drug loading inside of carrier nanoparticles, and a lack of applicability to a variety of medicinal agents [6]. Furthermore, residual precursor materials and excess organics present in unwashed or improperly formulated suspensions of nanoparticles may have detrimental effects on medical applications as well as toxic effects on the physiological system [7].

Considering the current limitations in nanomedical regulation, further standardization of the field must be completed to verify the feasibility and ensure the safety of nanomedicine. Advances in nanomaterial characterization through the enhancement of standard techniques, as well as the development of novel methods, must be made in order to drive the optimization of nanotechnologies for medical applications. Future efforts should focus on regulation and process control of nanoparticle synthesis and characterization methods, sustainability of nanoparticle formulations, and long-term toxicological effects of nanomedical-based products.

6.2.1 *Standardization of Nanomedicine Terminology*

The multidisciplinary nature of nanomedicine invites a similar multidisciplinary approach to descriptions and definitions. For example, while the academic materials science and engineering community largely views the term "nanophase" as referring to a special state of subdivision implying that particles or atomic clusters have average dimensions smaller than approximately 100 nm, many government groups promote a definition of nanoscale that includes sizes larger than 100 nm [8].

Some groups with a focus on the biological sciences have promoted an even larger concept of nanoscale of up to 1000 nm (1 μm) [9]. As another example, the term "carbon nanoparticle" has been used to describe a range of extremely diverse nanomaterials such as carbon-60, single-walled carbon nanotubes, and even diesel-based exhaust. In addition, some regulator definitions of "nanomaterial" integrate material characteristics other than size, such as composition, synthesis route, morphology, or biopersistance. The US Environmental Protection Agency (EPA) Office of Pollution Prevention and Toxics has stated, "In determining whether a nanoscale substance is a new or existing chemical, the agency intends to continue to apply its current inventory approaches based on molecular identity, rather than focus on physical attributes such as particle size" [10]. In addition, the Council of the European

Union has defined "nanomaterial" as "an insoluble or biopersistant and intentionally manufactured material with one or more external dimensions, or an internal structure, on the scale from 1 to 100 nm" [11].

6.2.2 *Standardization Organizations*

Various international standard-developing organizations have attempted to develop and implement terminology systems, characterization techniques, material specifications, and business processes to provide structure and clarity to the field of nanomedicine. A consensus between such groups on definitions, nomenclature, and standards for characterization and categorization of nanomaterials will facilitate the classification and regulation of the nanotechnology and nanomedical fields [2].

6.2.2.1 ASTM International Committee E56

The American Society for Testing and Materials (ASTM) International formed Committee E56 on Nanotechnology in October 2005 [12]. This committee addresses issues related to standards and guidance materials for nanotechnology and nanomaterials, as well as the coordination of existing ASTM standardization related to general nanotechnology applications. E56 has six technical subcommittees that maintain jurisdiction over the standards and guidance materials—E56.01 Informatics and Terminology; E56.02 Physical and Chemical Characterization; E56.03 Environment, Health, and Safety; E56.04 Intellectual Property Issues; E56.05 Liaison and International Cooperation; and E56.06 Nano-Enabled Consumer Products—in addition to E56.90 Executive and E56.91 Strategic Planning and Review [13].

The committee released its first standard, E-2456-06, Terminology for Nanotechnology, in 2006, which provides definitions for 13 terms specific to the industry [14]. For example, the prefix "nano" is defined in three ways: (1) SI units, (2) small "things," and (3) a set of concepts that must pertain to nanotechnology or nanoscience [9]. Some of the other terms defined in E-2456-06 include "nanotechnology," "nanoscale," and "nanostructured" [12].

E-2456-06 was developed by ASTM International in partnership with a variety of institutions, including the American Institute of Chemical Engineers, the American Society of Mechanical Engineers, the Institute of Electrical and Electronics Engineers, the Japanese National Institute of Advanced Industrial Science and Technology, NSF International, and the Semiconductor Equipment and Materials International [9]. The ASTM maintains that this collaboration will eliminate redundant resource allocation among a variety of standards organizations, provide for the pooling of technical experts in a single standards development venue, and ultimately assist in the creation of a global nanotechnology terminology framework [12].

6.2.2.2 ISO/TR 12802:2010

The International Organization for Standardization (ISO) is a worldwide federation of national standards bodies (ISO member bodies) [14]. In general, the work of preparing international standards is carried out through various technical committees. Draft international standards adopted by the technical committees are circulated to the ISO member bodies for voting [14]. Publication of an international standard requires approval by at least 75% of the ISO member bodies [14].

ISO/TR 12802 establishes core concepts for nanotechnology in a model taxonomic framework. It is intended to facilitate communication and promote common understanding of nanotechnology concepts. ISO/TR 12802 was prepared jointly by Technical Committee ISO/TC 229, Nanotechnologies, and Technical Committee IEC/TC 113, Nanotechnology Standardization for Electrical and Electronic Products and Systems [14]. Other vocabulary documents developed by ISO/TC 229 and IEC/TC 113 include the ISO/IEC 80004 series, which consists of various vocabulary sections [14].

6.2.3 *Standardization through Patent Systems*

The patenting of nanotechnologies and nanoenabled therapeutics also functions to regulate terminology and processes within the field of nanomedicine. As a component of the patent prosecution process, prior and/or related art is provided to the patent examiner. These cited pieces of art are published and made available to the public in the file history of the patent application. The cross referencing and publication of these documents creates a network of terminology in the nanotechnology and nanomedical fields.

6.2.4 *Physicochemical Characterization of Nanomedical Compounds and Devices*

The ISO has published a technical report providing guidance on the physicochemical characterization of manufactured nano-objects prior to toxicological assessment. TR 13014:2012, Nanotechnologies: Guidance on Physicochemical Characterization of Engineered Nanoscale Materials for Toxicologic Assessment, is intended to assist health scientists and experts to understand, plan, identify, and address relevant physicochemical characteristics of nano-objects before conducting toxicological tests on them [15].

6.3 Regulation of Nanomedicine

The use of nanomaterials in therapeutic and prophylactic agents, food additives, cosmetics, sunscreens, and other medical products has become widespread with minimal nanospecific oversight or direction from governmental agencies [1].

6.3.1 *US Food and Drug Administration*

The US Food and Drug Administration (FDA) maintains that the "unique size and properties of nano-scale materials do not warrant new regulation" and "the existing battery of pharmacotoxicity tests is probably adequate for most nanotechnology products" [16]. The FDA further asserts that its authorities are "generally comprehensive for products subject to premarket authorization requirements, such as drugs, biological products, devices, and food and color additives" [16].

Nevertheless, the FDA does participate in a variety of research programs with collaborating federal agencies, with the goal of better understanding the behavior of nanomaterials in biological systems.

In addition, the FDA collaborates with the National Nanotechnology Initiative (NNI), which serves as the central point of communication, cooperation, and collaboration for all federal agencies engaged in nanotechnology research [17].

The FDA has also funded a nanotechnology regulatory science program that further enhances the FDA's scientific capabilities [17]. The FDA's Nanocore facilities are located in Maryland as well as at the FDA's Jefferson Laboratories near Little Rock, Arkansas. The Arkansas Nanocore facility is a joint collaboration between the FDA's NCTR and the Office of Regulatory Affairs' Arkansas Regional Laboratory. Nanocore has been designed to support research scientists by providing the necessary equipment and educational materials to facilitate the characterization of nanomaterials [17]. Nanocore anticipates the needs of researchers through the development of novel methods to detect nanomaterials in biological samples following the use of nanomaterials in bioexperiments [17].

6.3.2 *US Environmental Protection Agency and US Consumer Product Safety Commission*

In 2014, the US EPA withdrew its proposal to regulate nanotechnology under the Toxic Substances Control Act (TSCA) and submitted an alternative rule for consideration and regulatory review by the Office of Management and Budget (OMB). The proposed reporting rule replaces strategies that had been under review by the OMB since 2010 [18].

The new rules eliminate the requirement on those parties intending to manufacture, import, or process designated nanomaterials for new uses to notify the EPA at least 90 days in advance, but retain the requirement that parties who manufacture nanoscale materials notify the EPA of certain information, including production volume; method of manufacture and processing, exposure, and release information; and available health and safety data. The reported information is intended to enable the EPA to evaluate the need for additional regulations under the TSCA [18].

The proposed rule will be published in the US Federal Register around March 2015. A period of public comment, EPA response, and additional OMB review will occur prior to the rule being finalized.

The EPA reporting rule could be the first major US federal rule to specifically regulate nanomaterials as a category.

The EPA and the US Consumer Product Safety Commission (CPSC) have entered into a collaborative agreement for worldwide research to articulate the potential impacts nanomaterials have on human health and the environment [19]. The collaborative research is part of a larger international effort with an emphasis on identifying nanomaterials, characterizing and quantifying the origins of nanomaterials, determining the ways in which nanomaterials interact with the human body and the environment, and sharing nanotechnology knowledge through online applications [19].

The CPSC is also collaborating with other US federal agencies to develop protocols to assess the potential release of nanomaterials from consumer products and rules for consumer product testing to better evaluate exposure [19].

6.3.3 *European Medicines Agency*

Nanomedicine is not specifically addressed in the European Union legislation on medicinal products and devices, tissue engineering, or other advanced therapies. To date, the European Medicines Agency's (EMA) existing regulations covering medical products and the extensive premarket safety assessment have ensured that the benefits of any nanoenabled therapeutic outweigh any identified risks or adverse side effects [20, 21].

Recommendations from the EMA's Committee for Medicinal Products for Human Use (CHMP) have led to the approval of a number of nanotechnology-based products [22]. Nanoenabled therapeutics in the form of liposomes, polymer–protein conjugates, polymeric substances, and suspensions have been provided marketing authorizations within the European Union under the existing regulatory framework (see, for example, Regulation 726/2004 on authorization and supervision of medicinal products for human and veterinary use, Directive 2001/83/EC on medicinal products for human use, Directive 93/42/EEC concerning medical devices, Directive 90/385/EEC relating to active implantable medical devices, and Directive 98/79/EC on in vitro diagnostic medical devices) [23–27].

The CHMP has initiated development of a series of reflection papers on nanomedicine in order to provide guidance to the developers of nanoenabled therapeutics [22]. These documents encompass the development both of new nanomedicines as well as nanosimilars (i.e., nanomedicines that are claimed to be similar to a reference nanomedicine) [22].

6.3.4 *Health Canada*

There are no currently no regulations specific to nanotechnology-based health and food products utilized by Health Canada. Rather, Health Canada has used existing legislative and regulatory frameworks to evaluate nanotechnology-based applications [28]. Various acts may be applicable to nanomaterials, including the Food and Drugs Act, the Canadian Environmental Protection Act of 1999, and the Hazardous Products Act [28]. Health Canada has recognized that new approaches may become necessary as the field of nanomedicine evolves.

Health Canada has adopted a broad definition for nanomaterials in the policy statement on Health Canada's working definition for nanomaterials [28]. The working definition enables Health Canada to establish internal inventories, request additional information, and integrate newly obtained knowledge into regulatory decision-making processes.

The Health Products and Food Branch (HPFB) of Health Canada participates in an interdepartmental Health Portfolio Nanotechnology Working Group that gathers information and acts as a discussion forum for issues related to nanomedicine. This working group contains members from Health Canada, the Public Health Agency of Canada (PHAC), and the Canadian Institutes of Health Research (CIHR) [28].

In addition, Health Canada participates in a number of international initiatives, such as the Working Party on Manufactured Nanomaterials of the Organisation for Economic Co-operation and Development (OECD) and Technical Committee 229 of the ISO, as well as collaborating with international counterparts [28].

6.4 Conclusion

Nanomedicine is a highly evolving, multidisciplinary technology spanning across many different industries, a fact that has caused the regulation of the field to become increasingly difficult. The continually developing state of environmental regulations, as governments and agencies attempt to gauge the potential impact of the manufacturing, use, and disposal of nanomaterials, creates further ambiguities about the practical implications and liabilities that developers of nanoenabled therapeutics may face.

Nevertheless, a recent survey by the OECD concluded that current regulations and legislation are adequate for the assessment of nanomedicines and other nanotechnology products. The survey of government bodies from 11 countries, Canada, Australia, Korea, Japan, Norway, Russian Federation, the U.S., France, Germany, the Netherlands, and Poland, as well as the European Commission concluded, "Foods and medical products that may contain nanomaterials or otherwise involve the application of nanotechnology are covered under existing national and/or regional legislative and regulatory frameworks" [29]. On the basis of the current stance on nanotechnology standardization and regulation the institutionalization of universal guidelines for nanomedicine is unlikely to occur quickly. Moreover, because the long-term impact of nanomaterials on the natural environment and human health is unknown and the multidisciplinary nature of the nanotechnology field, it is difficult to comprehensively regulate the technology in a single piece of legislation.

Given the increasing use of nanomaterials in medical applications, comprehensive legislation must soon be developed. All jurisdictions should continue to broaden legislation monitoring the development and approval of nanoenabled therapeutics. In addition, manufacturers should be required to research and report on the long-term effects of nanotechnologies. A mandatory safety reporting scheme should be introduced to monitor and memorialize the risks of nanomaterials present in medicines. These and other safety regulations would provide a level of protection for patients until sufficient research and nanospecific regulations can be imposed.

References

1. Lin, A. C. (2007) Size matters: regulating nanotechnology. *31 Harv Environ L Rev* **349**, 374.
2. Bowman, D. M., Hodge, G. A. (2007) A small matter of regulation: an international review of nanotechnology regulation. *8 Colum Sci Tech L Rev* **1**, 5.
3. Reynolds, G. H. (2003) Nanotechnology and regulatory policy: three futures. *17 Harv J Law Tech* **179**, 188–192.
4. Environmental Law Institute. (October 2005) *Securing the Promise of Nanotechnology: Is U.S. Environmental Law Up to the Job?*
5. Roy, I., Ohulchanskyy, T. Y., Pudavar, H. E., Bergey, E., Oseroff, A., Morgan, J., Dougherty, T., Prasad, P. (2003) Ceramic- based nanoparticles entrapping water-insoluble photosensitizing anticancer drugs: a novel drug-carrier system for photodynamic therapy. *J Am Chem Soc* **125**, 7860–7865.
6. Muller, R. H., Keck, C. M. (2004) Challenges and solutions for the delivery of biotech drugs: a review of drug nanocrystal technology and lipid nanoparticles. *J Biotechnol* **13**, 151–170.
7. Kumar, P., Mittal, K. L. (eds.) *(1999) Handbook of Microemulsion Science and Technology*, Marcel Dekker, New York.
8. Hackley, V. A., Ferraris, C. F. (2001) *The Use of Nomenclature in Dispersion Science and Technology, NIST Recommended Practice Guide*, Special Publication 960-3.
9. Klaessig, F., Marrapese, M., Abe, S. (2011) Current perspective in nanotechnology terminology and nomenclature, in *Nanotechnology Standards, Nanostructure Science and Technology*, Murashov, V., Howard, J. (eds.), Springer Science+Business Media.
10. Rizzuto, P., Pritchard, B. (2010) Industry developing nanoengineered goods frustrated by regulators' lack of definitions. *BNA Daily Environment Report* (May 17, 2010).
11. Jordan, W. (2010) EPA Office of Pesticide Programs, *Nanotechnology and Pesticides* (Apr. 29, 2010), http://www.epa.gov/pesticides/ppdc/2010/april2010/session1-nanotec.pdf.
12. ASTM International Tech News, http://www.astm.org/SNEWS/ NOVEMBER_2006/nano_nov06.html.
13. ASTM International, http://www.astm.org/COMMIT/SUBCOMMIT/E56.htm.

14. ISO/TR 12802:2010(en), *Nanotechnologies: Model Taxonomic Framework for Use in Developing Vocabularies; Core Concepts*, https://www.iso.org/obp/ui/#iso:std:iso:tr:12802:ed-1:v1:en.
15. TR 13014:2012, *Nanotechnologies: Guidance on Physicochemical Characterization of Engineered Nanoscale Materials for Toxicologic Assessment*, prepared by ISO Technical Committee (TC) 229 Working Group (WG) 3, Health, Safety, and Environmental Aspects of Nanotechnologies.
16. Food and Drug Administration, *FDA Regulation of Nanotechnology Products*, http://www.fda.gov/nanotechnology/regulation.html.
17. Nanocore, FDA Voice, http://blogs.fda.gov/fdavoice/index.php/tag/nanocore/.
18. OMB, Office of Information and Regulatory Affairs (OIRA), Executive Order Submissions Under Review. (2014) RIN: 2070-AJ54: EPA/OCSPP, Chemical Substances When Manufactured or Processed as Nanoscale Materials; TSCA Reporting and Recordkeeping Requirements.
19. EPA News Releases. (2012), http://yosemite.epa.gov/opa/admpress.nsf/cc70aa01e9532f3b852578d800648c5c/b3bdde177a3e57098525 7ad1006309d2!opendocument.
20. European Group on Ethics in Science and New Technologies (EGE). (2007) *Opinion 21: On the Ethical Aspects of Nanomedicine*, EGE.
21. European Medicines Agency. (2006) *European Medicines Agency: Reflection Paper on Nanotechnology-Based Medicinal Products for Human Use* (EMEA/CHMP/70769/2006), European Medicines Agency, London, U.K.
22. http://www.ema.europa.eu/ema/index.jsp?curl= pages/special_ topics/general/general_content_000345.jsp&mid=WC0b01ac05800baed9
23. European Parliament and Council of the European Union (EP and CEU). (2009) Regulation (EC) No 1223/2009 of the European Parliament and of the council of 30 November 2009 on cosmetic products(1). Official J Eur Union L 342 Volume 52 22 December 2009 ISSN 1725-2555 L 342/59 L 342/209.
24. Council of the European Communities. (1990) Council Directive 90/385/EEC of 20 June 1990 on the approximation of the laws of the member states relating to active implantable medical devices. *Official J L* **189**, 17–36.
25. Council of the European Communities. (1993) Council Directive 93/42/EEC of 14 June 1993 concerning medical devices. *Official J L* **169**, 1–43.

26. Council of the European Communities. (1998) Council Directive 98/24/EC of 7 April 1998 on the protection of the health and safety of workers from the risks related to chemical agents at work (fourteenth individual directive within the meaning of Article 16(1) of Directive 89/391/EEC). *Official J L* **131**, 23.
27. Council of the European Communities. (2001) Directive 2001/83/EC of the European Parliament and of the council of 6 November 2001 on the community code relating to medicinal products for human use. *Official J L* **311**, 67–128.
28. http://www.hc-sc.gc.ca/dhp-mps/nano-eng.php
29. http://www.theguardian.com/what-is-nano/putting-nanotechnology-regulation-under-the-microscope

Section III

Chapter 7

Overview of Intellectual Property Rights

Wim Helwegen[a] and Luca Escoffier[b,c]
[a] *Wim Helwegen Intellectual Asset and Legal Consulting*
[b] *Waseda University, Tokyo, Japan*
[c] *Innoventually S.r.l.s., Trieste, Italy*
wim@wimhelwegen.com, luca@innoventually.it

The most common instruments to secure investments in high-tech R&D are patents and trade secrets. Patents can be used to create a temporary monopoly on a certain invention, while trade secrets can be used to keep certain technologies or sensitive information concealed. In addition, database rights can be used to protect certain collections of research results from undue use. Other intellectual property rights (IPRs) such as trademarks are only relevant once the end product nears its market entry. This chapter will give an overview of the IPRs that are most relevant to the nanotechnology sector and other high-tech sectors. Due to their importance for nanotechnology, patents and trade secrets will be discussed more in detail than utility models, trademarks, design protection, and copyrights.

Commercializing Nanomedicine: Industrial Applications, Patents, and Ethics
Edited by Luca Escoffier, Mario Ganau, and Julielynn Wong

ISBN 978-981-4316-14-9 (Hardcover), 978-981-4613-14-9 (eBook)
www.panstanford.com

7.1 Patents

Patents are the most commonly used method of intellectual property (IP) protection in the high-tech sector. Under the European Patent Convention (EPC) only one kind of patents exists, whereas the US system encompasses several kinds of patents: utility patents, plant patents, and design patents. The patents discussed in this chapter are referred to as "utility patents" in the United States, unless stated otherwise.

A patent provides the proprietor with a negative right: the right to exclude. Depending on the jurisdiction in which the patent is valid, a patentee can exclude others from, for example, making, using, selling, leasing, importing, and keeping in stock the patented invention. In addition, the mere offer of any of the acts mentioned above could be forbidden by the patentee, again depending on the jurisdiction.

In no way does a patent confer a positive right on the patentee. If the authorities require regulatory permission to make, use, or market the invention, this permission should be sought separately from a patent.

Patents are normally valid for 20 years, although exceptions exist for patents on pharmaceutical products that have to be subjected to clinical trials. In the European Union (EU) such pharmaceutical patents can qualify for a Supplementary Protection Certificate, which provides additional protection after the expiry of the 20-year term of the patent. In the United States, a similar provision, known as Patent Term Extension, exists. These extensions are meant to compensate for the time during which the 20-year term of the patent has started but the patented invention could not be marketed due to the absence of regulatory permission.

7.1.1 *Requirements*

The right to exclude all others from using the patented invention can lead to a de facto monopoly for the patentee during the term of the patent. Because monopolies are undesirable from an economic perspective, patents are only granted if certain requirements are met. These requirements will be discussed briefly in this chapter and in more detail in Chapters 3 and 4.

Under the EPC, which is directly valid in 38 countries, the requirements for patentability are that the subject matter for which the patent is sought is an invention, susceptible of industrial application, which is novel and involves an inventive step (Article 51 (1) EPC). In the United States, patents can be granted to "whoever invents or discovers any new and useful process, machine, manufacture, or composition of matter, or any new and useful improvement thereof" (35 USC 101). The articles following Section 101 specify the requirement. Europe and the United States share the following criteria for patentability: (1) patentable subject matter, (2) novelty, (3) industrial applicability/utility, and (4) inventiveness/nonobviousness.

7.1.1.1 Patentable subject matter

Under the EPC, the first requirement is that the subject matter for which the patent is sought is an invention. The EPC does not provide a definition of "invention" but does name a number of categories that certainly do not qualify as inventions: discoveries, scientific theories, mathematical methods, aesthetic creations, schemes, rules, and methods for performing mental acts, playing games or doing business, and programs for computers and presentations of information (Article 52 (2) EPC). For the purpose of this chapter, the most important excluded category is "discoveries." As a general rule, one can say that a discovery is the finding of something that existed before. If, however, a practical use is found for a discovery, it might qualify as an invention.

In the United States, the invention or discovery should be a "process, machine, manufacture, or composition of matter." Courts have interpreted these four categories in such a broad sense that little seems to be excluded. It has been argued that "if you can name it, you can claim it" [1]. However, abstract ideas, laws of nature, and physical phenomena, such as the discovery of a plant, are not patentable subject matter [2].

Because one of the functions of patent law is to stimulate innovation and provoke new and better products, mere statements of facts on existing matter or methods are not patentable, whether in Europe or in the United States. To qualify for patentability, a practical use has to be added to the discovery.

7.1.1.2 Novelty

The next requirement is that the invention is new. The novelty requirement of the EPC is absolute. This means that "everything made available to the public by means of a written or oral description, by use, or in any other way, before the date of filing of the European patent application" is not regarded as novel (Article 54 EPC). The effect of this requirement is that every disclosure made to third parties not governed by a nondisclosure agreement destroys novelty and thus destroys patentability if the patent application has not been filed at the European Patent Office (EPO) before the disclosure. After the date of filing, you are free to disclose as much information as you wish. Even if you do not disclose anything yourself, the entire patent application will be published by the EPO 18 months after the date of filing.

The strict novelty requirement can be particularly burdensome in a technologically advanced and research-intensive area like nanotechnology. Submitting a paper that describes the invention to a magazine destroys novelty regardless of whether the paper is published or isn't—the magazine editor who reads the paper is also a third party unless that editor is under a nondisclosure agreement. The same goes for presenting inventions at a conference or in a business meeting where persons from outside the inventors' own organization are present. Even disclosing your invention to a friend or complete stranger can destroy novelty, although this risk can be theoretical because the burden to prove that such disclosure took place lies with the patent office or the person opposing or litigating against the patent. Nevertheless these risks should be avoided at any time.

US patent law can be a lot more lenient in case of the situations described above. In the United States, there is a so-called grace period of one year within which you can apply for a patent after having disclosed your own invention (35 USC 102 (b)). This leaves some time for evaluation after a publication or market entry.

Another difference with Europe is the method that is used to establish who is entitled to a patent in case two or more persons apply for a patent on the same invention. After the US patent system was reformed by the America Invents Act [3] in 2011, US patents are awarded to the "first inventor to file." Combined with the one-year grace period, this means that if you invent a product on January 1, publish it on March 1, and apply for a patent on May 1, you can still be granted a patent if another inventor invented the same product on February 1 and filed for a patent on April 1. In that case, your publication of March 1 destroys the novelty of your competitor's patent application. However, if your competitor files the patent application before your publication on March 1, then your competitor is the first inventor to file and will thus be granted the patent. Unlike under the previous US system, the fact that you are the first to invent said product is no longer relevant in these circumstances.

The advantage of the US system is that it leaves more room for trials and exercises less pressure on the inventor and the patent attorney. In Europe, the pressure to file the patent application as soon as possible is a lot higher. However, under the US system a lot of resources have to be dedicated toward proving that you are the first to invent. This requires precise administration because it can easily become the subject of a legal procedure. In Europe no such discussion exists, because the patent will be awarded to the first person to submit the application at the patent office, which can be proven with ease.

7.1.1.3 Industrial application/utility

If an invention can be "made or used in any kind of industry, including agriculture," it is deemed to have industrial applicability (Article 57 EPC). The term "industry" is interpreted broadly and includes "any physical activity of "technical character" [4]. The requirement mainly serves to exclude matter that is purely aesthetic or that cannot work because it operates in contradiction with the laws of nature, for example, a perpetual motion machine [5]. In US patent law, a fairly similar requirement exists, known as the utility requirement.

7.1.1.4 Inventive step/nonobviousness

The inventive step requirement stipulates that an invention must not be "obvious to a person skilled in the art" (Article 56 EPC). In the United States, where it is referred to as the nonobviousness requirement, no patent can be obtained if the invention "would be obvious at the time the invention was made to a person having ordinary skill in the art" (35 USC 103 (a)). The requirement's raison d'être is that society has no interest in granting exclusivity for inventions that do not advance a technology far enough to merit exclusivity.

The skilled person adds a subjective dimension to the inventive step requirement. The advantage is that it makes the requirement able to evolve with a technology. The knowledge of skilled persons increases over time, and so does the inventiveness threshold. The more prior art exists, whether patents, scientific papers, newspaper articles, or any other information, the more knowledge is attributed to the skilled person, and the more difficult it becomes to meet the requirement. The downside of the wide margin of appreciation that exists when determining the knowledge of the skilled person is that it creates a lot of space for disputes. Hence, the inventive step requirement plays an important role in most patent lawsuits.

7.1.1.5 Disclosure

Once the examiner of the EPO has established that the invention meets the aforementioned patentability requirements, there are several other requirements that the patent application has to meet. One of those requirements is the disclosure requirement. The EPC stipulates that the patent application shall disclose the invention "in a manner sufficiently clear and complete for it to be carried out by a person skilled in the art" (Article 83 EPC). In the United States, a similar requirement exists that requires the patent applicant to describe the invention "in such full, clear, concise, and exact terms as to enable any person skilled in the art to which it pertains [. . .] to make and use the same" (35 USC 112). In addition, an applicant for a

US patent also has to describe the best mode in which the invention can be made.

This requirement serves two main functions, an information function and a demarcation function. European and US patent applications become publicly available documents 18 months after the filing date. As a result, the inventor contributes the technical information that led to the invention to the public knowledge. The information should be clear enough to enable a skilled person to replicate the invention without undue burden. In this way, competitors who monitor the EPO or United States Patent and Trademark Office (USPTO) database can get an insight into your R&D process, but if this requirement would not exist, it would be possible to obtain a patent that is burdensome for society without giving society the benefit of the information contained in the patent. In return for this contribution to the common knowledge, the inventor receives the temporary right to exclude. In addition, patents have a demarcation function: you show in the patent what you claim as your property. If you own a plot of land and want to prevent people from entering it, you have to make clear somehow that the land is your private property, the same goes for patents. If the invention is not described in sufficiently clear terms, courts can invalidate the patent. Nevertheless, keeping the claims of a patent slightly ambiguous is attractive, and sometimes necessary, for a patent applicant. If a court has to decide whether the patent is infringed, it will look at the claims of the patent. The broader the claims, the more competing inventions can be held to infringe the patent. In addition, if the language of the patent is broad and ambiguous, competitors will get less insight into the applicants R&D process. Narrow and precise claims, however, may be too easy to circumvent and render a patent useless. It is understandable that most patentees aim for as broad as possible claims, but it is very important to keep in mind that the line between a valid broad claim and an insufficient disclosure can be extremely thin. Because an insufficient disclosure can lead to the invalidity of the entire patent, it is necessary to seek the assistance of highly specialized patent attorneys, as they are aware of all the nuances and pitfalls in your particular field of technology.

7.1.2 *Postgrant*

Once a patent is granted, it is entirely up to the patentee to decide how to use—or not to use—it.

Many patents end up in patent a company's patent portfolio, never to be looked at again. This is of course not the route to successful commercialization. To benefit from the patent, a myriad of possibilities exists. The patent can, for example, be sold (assignment), the right to make or use the invention can be "rented" out (license), the patent can be used as a collateral for debts, as a bargaining chip when being accused of infringement of another patent, and, of course, the patent can be used to exclude all others from making, using, selling, etc., the invention. To utilize the patent in the best way possible, it is essential to implement a patent—or IP—strategy well before the first patent is granted.

In the case of litigation, a big difference exists between Europe and the United States. Whereas a single US patent is valid for the entire United States, European patents (EPs) are often described as bundles of national patents. Accordingly, a US court has jurisdiction for the entire United States and a court within Europe only has jurisdiction for the country in which it is based. If an EP is infringed in multiple member-states, a separate infringement suit has to be filed in each member-state where the patentee wants to forbid the making, using, selling, etc., of the patented invention. These courts will apply their national patent legislation and legal precedents formed by case law within that country. As a result, it is possible that a product is ruled infringing in the UK, while the same product is ruled noninfringing in Germany. To counter differing outcomes of patent proceedings within the EU, a unitary patent (Unitary Patent Protection, UPP) and a unified patent court (UPC) are under development. If the legislative process proceeds as planned, applicants at the EPO can request that their patent be given unitary effect. Unlike the currently existing EPs, EPs with unitary effect are not subject to national formalities or payment of fees to national patent offices. UPP will automatically fall within the jurisdiction of the UPC. For EPs a transitional period of at least seven years will exist, during which applicants or proprietors can choose between national courts and the UPC. A ruling of the UPC is valid in all member-states. At the moment of writing it is assumed that all EU member-states except Croatia, Italy, Poland, and Spain will fully participate in UPP and the UPC.

7.1.3 *Exempted Uses*

Once the patent has been granted, not every possible use can be prohibited by the patent holder. In Europe, most countries have limited the reach of patents. In the United States, exemptions from patent infringement do exist but their scope is very narrow.

In the various national laws of Europe it is commonly accepted that patents only extend to commercial use. Commercial must be understood as meaning "nonprivate." As a result, the nonprivate use of a patented invention causes liability for infringement, even if the entity that uses the invention is a noncommercial entity such as a public hospital, a university, or a charitable organization. "Private" is interpreted narrowly and usually only applies to use within ones personal environment. In the United States, private—or personal—use is not exempted from infringement. "[W]hoever without authority makes, uses, offers to sell, or sells any patented invention, within the United States or imports into the United States" infringes that patent (35 USC 271). Whether the use is private or nonprivate is not an issue in terms of infringement.

In addition to the aforementioned private use exemptions, most European jurisdictions contain a research exemption. Research exemptions provide that acts that are conducted for experimental purposes "relating to the subject matter of the invention" cannot be held infringing. The exact definition of subject matter of the invention differs between member-states, but in general one can say that it is allowed to verify the claims of a patent or conduct research into, for example, improving the patented product as long as the patented product is not used outside the experimental context without the permission of the patent holder. Whether the experiment is conducted by a commercial or noncommercial entity is not relevant, neither is the intent of the research. If a company conducts experiments on a patented invention in order to develop a new product based on that invention, it can still rely on the research exemption. If the experiment is successful and a marketable product is developed in which the patented invention in included, the product can only be released with the permission of the patent holder, for example, via a license agreement.

US patent law also contains a research exemption. However, the exemption is that narrow that the practical use for industries and even universities is very limited. If an act is part of the legitimate business of an entity and it is not conducted "solely for amusement, to satisfy idle curiosity, or for strictly philosophical inquiry," the research exemption does not apply [6]. As a result of the legitimate business requirement, universities can also infringe a patent in the course of experimentation, because experimentation is part of their legitimate business.

7.2 Other forms of Intellectual Property

Beside patents, other different types of intellectual property rights (IPRs) might be envisioned for a more or less effective protection of nanotechnology innovations. Among them we can mention copyright, trademarks, integrated circuit layouts, designs, and a variant of patents, the utility model. Utility models are not foreseen in all jurisdictions and it consists of a right protecting something less than an invention, usually an improvement or betterment of an existing product or process, so it covers some sort of incremental invention. This kind of intellectual property right is also commonly referred to as "petty patent" or "small patent."

7.2.1 *Utility Models*

If a product has a lesser inventive character, a utility model can be a good method of obtaining protection. Utility model applications are generally not examined as thoroughly as patents and the requirements are usually lower, especially the inventiveness requirement. A utility model provides a shorter period of protection than the usual 20 years granted to ordinary patents. For instance, German utility models are valid for 10 years [7]. As to the possibility of protecting nanotech-related inventions through utility models, this could, for example, be useful in instances where the "minor" invention involves the employment of nanotech-related materials instead of a known traditional material. If the invention meets the less stringent requirements of a utility model and not those of a patent, the inventor will be able to protect the claimed invention through this residual kind of IP protection.

7.2.2 *Copyright*

Copyright can be used to protect the expression of an idea. Such expression can be a "production in the literary, scientific and artistic domain, whatever may be the mode or form of its expression, such as books, pamphlets and other writings; lectures, addresses, sermons and other works of the same nature; dramatic or dramatico-musical works; choreographic works and entertainments in dumb show; musical compositions with or without words; cinematographic works to which are assimilated works expressed by a process analogous to cinematography; works of drawing, painting, architecture, sculpture, engraving and lithography; photographic works to which are assimilated works expressed by a process analogous to photography; works of applied art; illustrations, maps, plans, sketches and three-dimensional works relative to geography, topography, architecture or science" (Article 2 Berne Convention for the Protection of Literary and Artistic Works).

So, as is easily inferable from the above definition, copyright covers most of human ingenuity's expressions. Because copyright can only be used to protect the expression of ideas in, for example, journal publications or conference presentations, it cannot be used to protect the idea itself. For protecting the ideas from being used by others, you can only rely on patents and to a lesser degree on utility models. For copyrights, no formal registration procedure exists. The protection arises as soon as the work is created and meets a certain standard of originality, which can differ between different jurisdictions. However, in the United States, it might be wise to register the copyright with the Copyright Office because such registration is necessary in case of a lawsuit. Copyright terms and the ways in which they are calculated differ between jurisdictions but are very long compared to patents. For example, in the EU, a copyright lasts for the life of the author plus 70 years. Contrary to popular belief, it is not necessary to use the © symbol in combination with the year of creation and name of the creator—or any other statement—in order to obtain a copyright. It can be useful, however, to use this statement in order to prove when the work was created.

7.2.2.1 Economic rights, moral rights, and other features

For sake of completeness, it is worth mentioning that there are economic rights and moral rights attached to copyrighted works. When we talk about the economic rights, we refer to all kinds of monetary benefits that can be related to the exploitation of copyrighted materials. Some examples are the licensing or assignment of the works. As to the moral rights, some jurisdictions recognize the right of the authors to be mentioned as the creators of the works and "to object to any distortion, mutilation or other modification of, or other derogatory action in relation to, the said work, which would be prejudicial to his honor or reputation" (Article 6bis Berne Convention). Moral rights are usually not alienable.

7.2.3 *Trademarks*

Trademarks are distinctive signs that are used to distinguish the products or services of an undertaking from those offered by others. Their main goal is to avoid confusion about the products origin among consumers. The word "sign" can be tricky sometimes, though. In fact, by "sign" usually trademark laws cover things like music, tridimensional shapes, and alike. To be more precise, in the EU, for example, Article 4 of the Community Trademark Regulation provides that "a community trademark may consist of any signs capable of being represented graphically, particularly words, including personal names, designs, letters, numerals, the shape of goods or of their packaging, provided that such signs are capable of distinguishing the goods or services of one undertaking from those of other undertakings."

So, for example, a trademark can easily protect a distinctive word or the peculiar shape of a bottle of water. For nanotechnology, the important thing to understand, though, is that trademarks do not protect the functional features of a product. These cannot be trademarked but should be patented instead. Unlike patents, which have a maximum validity of 20 years, trademarks have a potentially unlimited life span. As long as the trademark holder extends the mark in compliance with the applicable procedures, the trademark will continue to exist.

7.2.3.1 Requirements and characteristics

To get a trademark, the sign in question should be distinctive. By distinctive we mean that it must not be descriptive with regard to the product or service that you wish to trademark. So, for example, "Apple" is a distinctive sign for computers but not for a fruit seller. If a fruit seller would obtain a trademark to the word "Apple," this fruit seller could forbid all others from using the word "Apple" in relation to the sale of fruits. This, of course, would be highly undesirable. To prevent that established words are taken from the public domain and are claimed as private property, the distinctiveness requirement exists. When filing your trademark application, you have to designate for which class of goods and/or services you wish to register the trademark, for example, antibiotics, carbon, skin care, or transistors. These classes have been established in the Nice Agreement. The trademark office in Europe, either the national trademark office of the relevant member-state, or the Office for Harmonization of the Internal Market (OHIM) and in the United States, the USPTO, will determine whether the trademark is distinctive for the relevant product classes.

If the mark applied for is a "reproduction, an imitation, or a translation, liable to create confusion," of a well-known mark, both the European trademark offices and the USPTO will reject the application (Article 6bis Paris Convention for the Protection of Industrial Property). In case the mark applied for is identical to an existing but not well-known mark in the applicable product classes, the USPTO will reject the application, whereas in Europe it is the holder of the existing trademark who must file an opposition to the application trademark office where the application is filed. Trademark applications that are not identical but similar to existing trademarks have to be opposed by the right holder both in Europe and in the United States.

7.2.4 *Industrial Designs*

Industrial designs can be used to protect the appearance of a product but not its functional aspects. Because copyrights and trademarks can also be used to protect the appearance of products, overlaps of rights may occur. In fact, it is very common that products are protected by several independent IPRs at the same time. To be granted protection in Europe, the design must be novel and possess a distinctive character (Article 3 (2) Directive 98/71/EC). In the United States, design patents are available for products that are "new, original and ornamental" (35 USC 171). The ornamental, or decorative, character is important because the design does not protect functional features. Furthermore, US design patents can only protect products that have a utility in order to exclude purely aesthetic works from protection through design patents.

Like in copyrights and trademarks, the protection attached to nanotechnology can be solely indirect. In fact, let us assume that the unique appearance of a product is determined by the use of nanomaterials or nanotech-related process; in this case the product, where it meets the registrability requirements of a design, might well be protected, but what the protection will cover is exclusively the appearance itself. So let us say that a piece of furniture is made using carbon nanofibers and as a result its appearance can be protected as a design; in this case what the owner will be able to do is to enforce his or her rights against infringers who copy the appearance of the product but not the way the product is made.

7.3 Trade Secrets

Because nanotechnology can be hard to reverse engineer, trade secrecy may at present be an attractive form of protecting knowledge in the nanotechnology industry [8]. Trade secrets can be used to prevent that any kind of commercially valuable information is being made available to, or used by, others, but once the information has become public via an independent discovery, there is nothing that can be done against the use of that information. The way in which trade secrets are regulated varies between different jurisdictions, but the Agreement on Trade Related Intellectual Property Rights (TRIPs) sets minimum levels of protection that are applicable in every member-state (153) [9]. In the TRIPs context, a minimum level of protection means that member-states should provide at least the protection required by the TRIPs agreement but are allowed to impose further reaching protection for, say, trade secrets. Every TRIPs member-state must, at least, provide legal measures that enable the lawful possessor of information to prevent that information from being disclosed to, acquired by, or used by others if that information:

(a) is secret in the sense that it is not, as a body or in the precise configuration and assembly of its components, generally known among or readily accessible to persons within the circles that normally deal with the kind of information in question;
(b) has commercial value because it is secret; and
(c) has been subject to reasonable steps under the circumstances, by the person lawfully in control of the information, to keep it secret (Article 39 section 2 TRIPs).

Treating certain information as a trade secret has some obvious benefits over applying for patent protection for the possessor of that information. Unlike patents, that are subjected to a disclosure requirement, a trade secret, by its very nature, does not inform competitors about past or ongoing R&D activities. Furthermore, there is no formal procedure that one has to follow to obtain a trade secret and the subject matter to which it can be applied is much broader than patents. Patents are only available for restricted categories of inventions, but trade secrecy can be applied to any information that meets the requirements of the TRIPs agreement, although the exact way in which these requirements are applied differs between different jurisdictions. If it is questionable whether an invention will withstand the examination proceeding at the patent office, trade secrecy might be an option. A major advantage of trade secrets over patents is that the duration of a trade secret is potentially unlimited, unlike patents, which are restricted to a

20-year term. The downside of the undefined term is that there is no certainty on how long the trade secret can be maintained. If a competitor independently obtains the same information after, say, one year, a patent would have been a better way of protecting the invention. A patent could in such case be used to forbid the competitor to use the invention, whereas a trade secret loses its entire value as soon as it publicly available. If the invention is of such a complexity that it would take competitors more than 20 years to develop it independently, trade secrecy might be more attractive.

The aforementioned lack of a formal procedure does certainly not mean that trade secrets are easier to obtain and maintain than patents. To make sure that the information remains a secret, a strict secrecy policy has to be applied and detailed records have to be kept. To a certain extent this overlaps with the secrecy and record keeping that is required to obtain patent rights.

References

1. Adelman, M. J., Rader, R., Thomas, J. (2009) *Cases and Materials on Patent Law*, 58, Thomson Reuters.
2. Ibid.
3. 112th Congress Public Law 29 Leahy-Smith America Invents Act.
4. *Guidelines for Examination in the European Patent Office*, Part C, Chapter IV, 5.1.
5. Ibid.
6. 307 F.3d 1351 (Fed Cir. 2002).
7. §23 section 1 Gebrauchsmustergesetz.
8. Lemley, M. A. (2005) Patenting nanotechnology. *Stanford Law Rev* **58**, 19, http://ssrn.com/abstract=741326 or doi:10.2139/ssrn.741326.
9. http://www.wto.org/english/thewto_e/whatis_e/tif_e/org6_e.htm

Chapter 8

IP Valuation: Principles and Applications in the Nanotechnology Industry

Efrat Kasznik
Foresight Valuation Group, Palo Alto & Stanford Graduate School of Business, USA
ekasznik@foresightvaluation.com

8.1 Overview of IP Valuation

8.1.1 *What Is Intellectual Property Valuation?*

8.1.1.1 Brief history of IP valuation in the United States

With its origins in the intellectual property (IP) litigation of the 1980s and 1990s, the valuation of IP (primarily patents) in the United States was initially limited to damages calculations in legal cases involving claims such as patent infringement. With the introduction of tax planning involving IP, such as transfer pricing and patent donations, the valuation of intangibles became critical in nonlitigation circumstances as well. Companies were required to include in their tax reporting the fair market value (FMV) of IP involved in transactions, such as the intercompany transfer of IP or the donation of a patent to a university. New accounting rules related to business combinations in the United States, introduced

Commercializing Nanomedicine: Industrial Applications, Patents, and Ethics
Edited by Luca Escoffier, Mario Ganau, and Julielynn Wong

ISBN 978-981-4316-14-9 (Hardcover), 978-981-4613-14-9 (eBook)
www.panstanford.com

in the early 2000s, expanded the need for IP valuations even more, as companies were now required to report the fair value (FV) of intangibles that were purchased with a target in a mergers and acquisitions (M&A) deal. These *compliance* situations—litigation, accounting, and tax reporting—carry with them a high degree of scrutiny by the court or regulating authorities and require a third-party, IP valuation expert's opinion in the form of a report or testimony.

In parallel to the proliferation of tax and accounting rules, which mandated the valuation of IP in certain transactions, around the same time (late 1990s–early 2000s) the field of intellectual asset management (IAM) was starting to gain momentum with US corporations. Large companies with significant patent portfolios (like IBM, Dow, and others) were leading the way, and with the increase in sophistication of active IP portfolio management, came the need for valuation. The types of activities where a valuation became increasingly important include spin-offs, in-kind contributions, licensing, patents sales, and other commercialization activities. Since many of these activities are not always subject to tax or accounting reporting, these transactions can be referred to as *noncompliance* activities. In these situations, due to the low to medium degree of scrutiny and the lack of reporting requirement, the valuation is often done in-house or between the parties without the involvement of a third-party IP valuation expert.

8.1.1.2 IP valuation methodologies

There are three common methods for valuing intangible assets: the *market method*, the *income method*, and the *cost method*. These valuation methodologies were largely borrowed from the methodologies applied in the valuation of tangible assets (like real estate, machines, inventories, etc.). As a result, they are more suitable for tangible assets and some of them are challenging to implement when intangibles are involved. Each of these valuation methods is briefly described below.

Market methods are useful where there is a market demand for an intangible asset and there are other similar intangibles that have traded hands under specified market conditions. In these circumstances, the market price for comparable assets may be used (with appropriate adjustments) as a measure of the market value of the intangibles at issue. The market method of valuation relies upon the availability of comparable external transaction data, which are sometimes difficult to find. In practice, although it is the preferred approach for many assets, it is often not possible to use market approaches if there is no observable active market for the intangible asset. The market approach is therefore not commonly applied.

Income methods are the most commonly used methodologies in the valuation of intangibles. Income-based approaches involve calculating the value of an intangible asset on the basis of the aggregate income stream that ownership of such intangible asset will provide. That income stream, net of any costs associated with its production, is discounted to its net present value (NPV) to determine the value of the intangible asset. The application of such methods requires the projection of economic income that is directly generated by the asset over its economic life. These projections are converted into the NPV by using a present value discount rate, which represents the required rate of return over the intangible asset.

Cost methods involve an analysis of all cost components that went into creating the intangible asset, such as materials, labor, and overhead. Cost methods are hardly used when valuing intangible assets, primarily because these methods do not consider future economic benefits arising from the asset. The application of this approach is only appropriate for assets that are usually accounted for by the cost of reproduction, such as software and assembled workforce.

8.1.1.3 Types of intangible assets

The types of intangible assets that are covered in this chapter fall into several subcategories. One very useful framework for classifying intangibles can be found in the United States: generally accepted accounting principles (GAAP) definitions as found in Accounting Standards Codification (ASC) 805. The guidelines of

ASC 805 are used for identifying intangibles involved in business combinations, such as M&A transactions [1].

Among the ASC 805 intangible asset categories, the most common groups of assets found in the nanotechnology industry are the following:

(a) Technology-based intangible assets: Innovations and technological advances that are protected by contractual or legal rights. This group includes the following assets:
- Patented and unpatented technology
- Computer software and mask works
- Databases, including plants
- Trade secrets, such as secret formulae, processes, recipes, etc.

(b) Marketing-related intangible assets: Assets that provide value to the marketing or promotion of products and services. This group includes the following assets:
- Trademarks, trade names, service marks, collective marks, certification marks
- Trade dress (unique color, shape, or package design)
- Internet domain names
- Noncompetition agreements

(c) Customer-related intangible assets: A customer relationship exists between an entity and its customer if the entity has information and regular contact with the customer and if it stands to benefit from future contracts that are reasonably anticipated from that customer. This group includes the following assets:
- Customer lists
- Order or production backlog
- Customer contracts and relationships

8.1.2 *IP Valuation Standards in the United States*

8.1.2.1 IP valuation landscape in the United States

Most IP valuations in the United States are done in *compliance* situations, for financial reporting, tax compliance or litigation damages. There is a fundamental difference between the IP valuation requirements in compliance versus noncompliance situations. In compliance situations the valuation is mandatory and is usually done after the deal has already been finalized. There is a single point value that needs to be recorded (as opposed to a *range* of values that needs to be negotiated). On the other hand, most IP transactions today are done in noncompliance situations, where there is no mandatory reporting requirement. If an IP valuation is conducted under such circumstances, it is not reported nor regulated under any standards.

Valuations done in noncompliance situations are usually done in-house for purposes of negotiations, and therefore there would not be a single point value but rather a range of values that needs to be negotiated between a buyer and a seller. In recent years, a large volume of IP transactions involving the sales of patents were carried out through IP brokers and various kinds of IP funds (such as patent aggregators, defense funds, etc). There are rarely any formal IP valuations done in conjunction with these transactions, as most of them rely on legal claim chart analysis and some heuristic rules of thumb as to the discounts that should be applied to future cash flows.

We turn next to discuss in more detail the IP valuation activities under the most common compliance situations: litigation damages, financial reporting (accounting), and tax reporting.

8.1.2.2 IP valuation for litigation damages

The litigation of intellectual property in the United States has seen a sharp increase since the 1990s, both in the number of cases filed annually and in the cost of litigation. According to the American Intellectual Property Lawyers Association (AIPLA), the number of patent infringement cases has doubled in 10 years—from about 1700 in 1995 to over 3300 in 2005 [2]. The number has since leveled somewhat but is still significantly higher than a decade ago. The cost of patent litigation has sky rocketed as well; the average cost of litigating a patent case through trial, according to the AIPLA, is estimated at \$1–\$3 million, depending on the amount of damages and the size of the case.

The field of nanotechnology is at its early stages of patenting, and litigation activity has so far been limited. The first case filed in a potential wave of patent infringement litigations based on nanotechnology patents, *DuPont Air Products Nanomaterials v. Cabot Microelectronics* (filed in Jan. 2007), represents what many feel will turn out to be a growing trend amongst companies. While it is difficult to predict how the litigation environment will evolve to absorb the changes brought by the nanotechnology field, all eyes will be on the top players and how they react to legal challenges related to their IP portfolio.

IP valuation analysis is implemented in IP litigation situations when it comes to damages calculations. The purpose of damages is to make the plaintiff whole, that is, compensate the injured party such that it would return to the position it would have been in *but for* the infringement of its intellectual property. Generally speaking, there are four types of IP litigation in the United States: patents, copyrights, trademarks/trade dress, and trade secrets. Below is a brief overview of the damages calculation standards for patent litigation [3].

When the plaintiff can successfully show infringement of a valid US patent, 35 USC §284 states that "upon finding for the claimant the court shall award the claimant damages adequate to compensate for the infringement but in no event less than a reasonable royalty for the use made of the invention by the infringer, together with interest and costs as fixed by the court." The case law implementation of this statute allows for patent infringement damages based on lost profits, a reasonable royalty or a combination of both depending on the circumstances of the case.

Much legal guidance, case laws, and precedents exist in the United States as to how patent infringement damages are calculated. The first step in quantifying damages in a patent infringement matter involves the application of a four-part test set forth in *Panduit Corp v. Stahlin Bros Fibre Works*, 575 F2d 1152 (6th Cir 1978). In the *Panduit* case, the court found that the conditions necessary for the calculation and recovery of lost profits are the following: a demand for the patented product, the absence of noninfringing alternatives, the existence of sufficient manufacturing and marketing capacity on the part of the plaintiff, and the plaintiff's ability to quantify lost profits.

Should the *Panduit* conditions not apply, the damages calculations revert to reasonable royalty. The determination of reasonable royalty damages involves the construction of a hypothetical negotiation scenario between the parties at the date of the infringement. Such an analysis is generally conducted through a consideration of the 15 factors set forth in *Georgia-Pacific Corp v. United States Plywood Corp*, 318 F Supp 1116, 1120 (SDNY 1970). These are economic factors that increase or decrease the reasonable royalty rate that would have been the result of the hypothetical negotiations.

8.1.2.2.1 Trademark, copyright, and trade secret damages

Damages in trademark, copyright, or trade secrets cases are estimated using a broader range of measures as compared to patent infringement damages. For example, the disgorgement of the defendant's profits may be an allowable measure of damages in these types of cases (when such damages are not allowed in patent cases). Copyright infringement offers an additional approach to damages not available in other IP matters—statutory damages that are set at a fixed amount of dollar per infringement.

8.1.2.3 IP valuation for financial reporting

According to US GAAP rules, intangible assets are only reported on the financial statements when they are paid for in a business combination transaction, such as an acquisition. As a result, "home grown" IP assets, like patents and trademarks, are not measured or reported on the balance sheet of the company that created them. However, if that company buys another company that owns patents and trademarks, the acquired intangible assets of the target company will be valued at their FV and reported on the acquiring company's books. That is an interesting anomaly that is frequently pointed out by members of the IP community; however, it is unlikely that any changes to that accounting treatment will take place in the near future, primarily because of the conservative nature of GAAP rules, and the somewhat speculative nature of IP asset valuations.

As long as these assets are not priced by a market transaction, their value is not certain enough for financial reporting purposes.

That being the case, the valuation of IP for accounting purposes is primarily done in the context of business combinations (such as M&A deals) and is governed by GAAP pronouncements such as ASC 805 (first introduced in 2001 as Statement of Financial Accounting Standards 141). The valuation is done as part of the overall process of purchase price allocation (PPA), when the deal price paid for the acquired company is allocated among the various assets that are comprised in that company. All assets of the target company, tangible and intangible, as well as its liabilities, are identified and valued at their current FV. When valuing assets at FV, the appraisal needs to assume their "highest and best use," which refers to the use of an asset by market participants that would maximize the value of the asset, even if the intended use of the asset by the holding entity is different.

The standard of valuation applied in business combinations is FV, defined as the price that would be received when selling an asset (or paid when transferring a liability) in an orderly transaction between market participants. "Market participants" are defined as buyers and seller in the principal market for the asset that have all the following characteristics:

a. Independent and not related parties
b. Knowledgeable, having a reasonable understanding about the asset (or liability) and the transaction on the basis of all available information, including information that might be obtained through normal and customary due-diligence efforts
c. Able to transact for the asset or liability
d. Willing to transact for the asset or liability, that is, motivated but not forced or otherwise compelled to do so [4]

The FV of all assets (net of liabilities) is then compared to the acquisitions price, and the residual amount is recorded as *goodwill*, which by itself is considered an intangible asset. Goodwill represents the future economic benefits arising from other assets acquired in a business combination that are not individually identified and separately valued. The results of the PPA analysis, including goodwill, are then reported in the new, combined financial statements of the two companies. In the following years, goodwill is tested annually for impairment, a process that involves a valuation analysis. If the test shows indication of impairment, then some of the intangible assets' FV estimates may be revisited, and certain assets, or goodwill, may be written off, as necessary.

The US GAAP valuation guidelines state a preference for market-based valuation methods when market data is available. In reality, the markets for intangibles are very thin, the assets are uniquely different from each other, and most transactions are not publically reported. As a result, most IP valuations done for accounting purposes rely on income-based approaches. In a recent study done by the global accounting firm KPMG it was reported that income-based valuation methodologies are the most commonly used for FV IP valuation [5]. The study looked at over 300 M&A transactions between 2003 and 2007 by industry groups. The results of the study show that the percentage of the acquisition purchase price allocated to intangible asset (based on the FV valuation analysis) varies significantly across industries. The study further determines that in the majority of the industries analyzed, the percentage allocation of the purchase price to goodwill is typically over 50%.

8.1.2.4 IP valuation for tax reporting

The most common application of IP valuation in tax reporting and planning is related to *transfer pricing* between affiliated companies. The intangible nature of IP rights renders them easily movable and creates the potential for valuable tax planning. This often involves the migration of IP into favorable tax jurisdictions or into IP-holding companies [3]. The role of the IP valuation expert is to calculate the FMV of the IP portfolio when it is the subject of an intercompany transfer, such as the migration to another country. The IP valuation expert can also be asked to determine the *arm's length* royalty rates for licensing IP rights between affiliated companies; these are the royalty rates that approximate the rates that would have been established between two unrelated parties.

Due to the complexity of global tax planning, the Internal Revenue Service (IRS) in the United States established a comprehensive set of transfer pricing compliance and documentation requirements. IRS Section 482, and its accompanying regulations, lays out the general compliance framework for transfer pricing schemes involving US companies, including the list of acceptable valuation methodologies for transfer pricing situations. These methodologies generally fall into three categories [3]:

1. *Transaction-based methods*: The arm's length intercompany royalties are determined on the basis of the terms of *comparable* licensing deals between nonaffiliated companies.
2. *Profit-based methods*: The arm's length intercompany royalty rates are determined on the basis of the relative contribution of the intangible asset to the overall combined company's profit margin.
3. *Cost-sharing methods*: Cost-sharing agreements are used to divide the cost of developing intangible assets between affiliated parties, based on their respective return from the future assets. The IRS sets out a list of requirements for cost-sharing agreements' qualifications, including specific documentation and reporting rules.

8.1.3 *IP Valuation Circumstances in Europe*

8.1.3.1 Litigation damages

The laws of IP damages in Europe also rely on the premise of restoring the injured party to the position it would have been in *but for* the infringement of its intellectual property, although the legal procedures could vary from the United States. For example, in the United Kingdom, damages hearings constitute a separate trial, which only kicks in once liability has been established; damages hearings are therefore less frequent than in the United States as the parties have a greater motivation to settle [6].

The concept of reasonable royalty damages, based on a hypothetical licensing negotiation between the IP owner and the infringer, is common in Europe as well. The courts apply a range of royalty determining considerations, similar to Georgia-Pacific, although European courts have not clearly established a set of relevant criteria yet.

8.1.3.2 Financial reporting

As a result of the international harmonization of accounting standards, the treatment of IP valuation in business combinations is similar in Europe and in the United States. All member states of the European Union (EU) are required to use the International Financial Reporting Standards (IFRS) system of accounting standards, as adopted by the EU for listed companies since 2005. The United States is expected to follow suit in the next few years and transition into the IFRS from the current US GAAP system. IFRS 3 is the standard that applies to valuation for business combinations (the counterpart of ASC 805 in the United States).

Interestingly enough, it should be noted that prior to the EU adoption of IFRS standards in 2005, the United Kingdom was one of the only countries where home-grown intangibles (i.e., intangibles that were internally developed by the company) could be measured and presented as assets on the balance sheet. That is no longer the case as IFRS rules mostly recognize intangibles as assets for accounting purposes if they were purchased in a business combination.

8.1.3.3 Tax reporting

Europe is a major hub of transfer pricing activity, as several European countries are favorite locations for setting up offshore IP-holding companies. Currently, Switzerland and Ireland are the preferred locations for US-headquartered corporations, while EU-based corporations tend to prefer Ireland and Luxembourg. Switzerland, Ireland, and Luxembourg have good treaty networks and very advantageous tax rates. For instance, the tax rate in Switzerland is determined by negotiation with the local canton and is usually in the range of 4% to 8% [7].

The valuation of IP assets is critical at the time of transfer of the IP portfolio into the IP-holding company and needs to be done by a third-party, independent valuation expert. Each IP asset should be separately assessed. The IP-holding company needs to have to function as a real business; otherwise it might be treated as a "controlled foreign corporation," in which case its accounts would have to be consolidated with those of the parent company and it would lose the benefits of operating in a low-tax jurisdiction. To overcome this problem in the EU, the IP-holding company has to have "functionality," which is demonstrated by a staff that actively manages the IP portfolio.

8.2 The Application of IP Valuation in the Nanotechnology Industry

8.2.1 *IP Valuation vs. Evaluation*

When dealing with products that are years away from commercialization, there is no easy way to model cash flows for valuation. Many assumptions need to be made on the probability and success of future events, sometimes in the very far future. Some of the assumptions are related to risk, such as technology risk of development, risk of market acceptance, or regulatory risk. One common way for dealing with risk is conducting an *evaluation* of the technology or IP assets prior to the actual *valuation*. An evaluation involves a nonquantitative assessment of factors that are specific to the technology (development, market, regulatory) or the IP (quality of patents). A valuation involves the assignment of a precise dollar value to an IP portfolio, which takes into consideration the results of the evaluation process.

In emerging industries, like nanotechnology, the need for an evaluation process is becoming a critical part of the process of assigning a value to a patent portfolio, or any other IP asset, because any dollar value is subject to all the risk factors involved in the commercialization, regulation, and market adoption. It is therefore common to apply techniques in which a *quantitative* assessment (the valuation) is coupled with a *qualitative* one (the evaluation) in order to achieve a monetary valuation that can take into consideration all the variables that characterize the development and commercialization of nanotechnology products and processes. When a precise dollar valuation of IP assets is required, a thorough evaluation process can be the key to improving the accuracy of that number.

8.2.2 *Valuing an IP portfolio in the Nanotechnology Industry*

When it comes to creating, managing, and commercializing a nanotechnology IP portfolio, there are some common themes related to IP portfolio management than run throughout the industry.

Three of these IP management topics stand out as broadly applicable to a wide range of nanotechnology IP holders around the world, as the industry follows the path from early stage to maturity:

- Patenting along the nanotechnology value chain
- Technology transfer from university to industry
- Mitigating litigation risk

The following is a brief discussion of each of these topics, including a discussion of the role that IP valuation and evaluation can play in supporting these challenges.

8.2.2.1 Patenting along the value chain

The concept of the *nanotechnology value chain* is closely related to the interdisciplinary nature of the industry. This concept is illustrated in a 2003 *Intellectual Asset Management* (IAM) magazine article [8]. The authors claim that the key to a strong IP portfolio is found in patenting across the entire nanotechnology value chain, taking into consideration the following elements:

- The basic chemical composition
- The physical structure
- The tool for developing that structure
- The method or process for using that tool
- The article of manufacture (end product)

Vertical integration along the supply chain can be done by either licensing or acquisitions. Both types of transactions require the support of IP valuation as well as evaluation analysis.

Licensing or acquisitions include a long process of technology and IP due diligence prior to the transaction, where the IP in particular is evaluated for quality and robustness. Once the evaluation process has concluded, the buyer or licensee would need to further engage in valuation analysis to determine the terms of the license agreement, or to value the IP assets as part of the acquired target.

8.2.2.2 Technology transfer from university to industry

Nanotechnology relies heavily on federal funding, technology transfer, and university inventions. The transfer of new technology from university laboratories to the private sector has a long history and has taken many different forms. Prior to the 1980 enactment of the Bayh–Dole Act, companies did not have exclusive rights under government patents to manufacture and sell resulting products [9]. As a result, companies were reluctant to develop new products when competitors could also acquire licenses to the same technology. The government remained unsuccessful in attracting private industry to license government-owned patents. The Bayh–Dole Act permitted universities and small businesses to elect ownership of inventions made under federal funding and to become directly involved in the commercialization process. This policy also allows for exclusive licensing when combined with diligent development and transfer of an invention to the marketplace for the public good.

With the passage of the Bayh–Dole Act, colleges and universities immediately began to develop and strengthen the internal expertise needed to effectively engage in the patenting and licensing of inventions. In many cases, institutions that had not been active in this area began to establish entirely new technology transfer offices, building teams with legal, business, and scientific backgrounds. As a result, many new technologies have been diligently and successfully introduced into public use. Another significant result of the Bayh–Dole Act is that it provides a strong incentive for university–industry research collaborations.

The technology transfer process from university to industry is primarily done in two ways: licensing and joint ventures (JVs). In both instances, there is a thorough evaluation process of the IP that is required to understand the specific encumbrances that usually accompany IP originating from universities. Licensing terms can be limited in scope and time. Ownership rights are sometimes ambiguous, in particular ownership in future IP that relies on the original IP that is being licensed or contributed in the original transactions.

The IP evaluation is followed by an IP valuation analysis that needs to take place in order to support the execution of the transaction. In licensing, deals, the parties need to agree on the royalty rate, royalty terms (lump sum, running royalty, or a combination of both), and other terms related to the license. In JVs, a university would typically contribute the IP, while the industry partner contributes the tangible assets (usually cash or equipment). To figure out what share of the venture each side is entitled to, the university's "in kind" contribution of IP needs to be valued, a task that is usually assigned to a third-party valuation expert.

8.2.2.3 Mitigating litigation risk

The biotechnology and semiconductor industries experienced accelerated patenting activity, followed by litigation, over the last 20 years [10]. This pattern could emerge in the nanotechnology field, especially in light of the claim overlap due to the unusually high number of patent applications covering similar technologies. Bringing a product to market involves navigating patent thickets, and facing potential litigation down the road as the industry matures seems highly likely.

Legal experts following the nanotechnology industry vary in their assessment of the future of litigation in the industry. Some believe that cross licensing is an effective tool for resolving patent thickets. The argument here is that by carefully carving the fields of use in a cross-licensing deal, each party can have an exclusive field of use that does not overlap with the other party [11]. The 2005 Lux research study goes on to predict that the nanotechnology industry will, for the most part, manage to avoid a "self destructive IP war" through a flood of cross-licensing agreements and other types of IP licensing.

On the other hand, other legal experts are not as optimistic as to the mitigating role of cross licensing when it comes to future patent litigation. Their argument is that cross licensing has worked in the semiconductor industry due to the oligopolistic nature of that industry, where there are a relatively small number of large firms with similar products and similar IP portfolios [12]. The nanotechnology industry is inherently different: there are many patent holders in a variety of different industries. Two start-ups trying to commercialize the same product in the same market are more likely to try and litigate each other out of business, as opposed to cross licensing, as has been the case in the biotech industry.

IP evaluation takes center stage in enforcement situations. The quality and scope of the claims as well as other characteristics of the patent at issue and its relationship to products in the market (evidence or indication of use) are the main drivers of patent value in the context of litigation. This is one example where the evaluation is driving the valuation of the IP assets. Often times there are no formal valuations done until the time of trial, where damages are calculated by economic experts. The parties rely on evaluations done by technical and legal experts to determine their litigation strategy.

References

1. Chapter B4, in *Financial Reporting Developments, Business Combinations: Accounting Standards Codification 805*, revised October 2010, Ernst and Young.
2. *Cost of Patent Litigation*, AIPLA Mid-Winter Conference, Jan. 25, 2008, http://www.aipla.org/Content/ContentGroups/Speaker_Papers/Mid-Winter1/20083/Showalter-slides.pdf.
3. Tanger, C. D., Palim, M., Davis, J., Samuel, T. *Accounting for Damages in Intellectual Property Litigation,* PwC, Atlanta, Washington DC, London and Sydney.
4. All definitions are taken from the ASC Master Glossary.
5. *Intangible Assets and Goodwill in the Context of Business Combinations, An Industry Study*, KPMG: Corporate Finance Advisory, May 2009.
6. Anson, W. (2006) Chapter 18, in *Fundamentals of Intellectual Property Valuation: A Primer for Identifying and Determining Value*, American Bar Association.
7. Cohen, L., Finn, S. (2010) *Offshoring Your IP holdings*, New Legal Review.
8. Wild, J. (2003) Patent challenges for nanotech investors, *Intellectual Asset Management*, September/October 2003 (sidebar titled "Developing the Nanotech Patent Portfolio" by Rachel E. Schwartz and John E. Cronin of the ipCapital Group, Inc.).

9. University of California Office of Technology Transfer, *The Bayh-Dole Act: A Guide to the Law and Implementing Regulations*, http://www.ucop.edu/ott/faculty/bayh.html.
10. *Nanotechnology: Thinking Small in a Big Way* (2008), JonesDay.
11. Maebius, S. B., Radomsky, L. (2005) *The Nanotech IP Landscape: Increasing Patent Thickets Will Drive Cross-licensing*, Foley and Lardner.
12. Serrato, R., Herman, K., Douglas, C. (2005) The nanotech intellectual property ("IP") landscape. *Nanotechnol Law Business J*, 2(2), Article 3.

Chapter 9

Commercialization, Valuation, and Evaluation of Nanotech Innovations[a]

Luca Escoffier
Waseda University, Tokyo, Japan
Innoventually S.r.l.s., Trieste, Italy
luca@innoventually.it

9.1 The Commercialization of Nanotechnology Innovations

Nanotechnology commercialization embodies the paradigm of the transfer of innovation from research institutions. So far, it has been mentioned how critical is the role of universities in the development of nanotechnologies [1, 2]. Kesan [3], in his recent study on the information submitted from 1996 to 2003 by 94 US universities to the Association of University Technology Managers (AUTM), concludes that the most frequent practice consists of licensing out

[a]The present chapter is partly reproducing the content of "Reinterpreting Patent Valuation and Evaluation: The Tricky World of Nanotechnology," which is reprinted with permission from *European Journal of Risk Regulation*, **2**(1) 67–78, 2011, Copyright © Lexxion Publisher, Berlin.

Commercializing Nanomedicine: Industrial Applications, Patents, and Ethics
Edited by Luca Escoffier, Mario Ganau, and Julielynn Wong

ISBN 978-981-4316-14-9 (Hardcover), 978-981-4613-14-9 (eBook)
www.panstanford.com

universities' inventions for recouping the legal expenses incurred for patenting activities, and, more, in general, pursuing a revenue-centric approach. He then concludes that universities should rather think of following different routes like actively pursuing commercialization activities, adopt an open collaboration scheme, and adopt royalty-free licensing (for a detailed analysis of nanotechnology's potential and technology transfer strategies in Finland, see Ref. [2]). Bastani et al. [4], for example, advocate the great importance of universities in nanotechnology development, and, consequently, of effective technology transfer initiatives, and suggest alternative ways to collaborate by fostering the industry–academia relationship, like the negotiation of partnership agreements right after the publication of invention disclosures. In fact, in this field, companies may realize that a solution to a technical problem and its industrialization may need further refinement and research, which can be done in parallel with the relevant university. Stewart [5] argues that there is a difference between biotech ventures and nanotechnology ones and suggests some alternative ways to license and commercialize nanotech-based innovations from the university setting. He advocates the value of nanotechnology as a platform technology and argues that university spin-offs should focus on one application and exclude the others, at least at the beginning of the development.

So far, many attempts have been made to highlight the most common barriers to the commercialization of nanotechnology-related innovations [6]. For example, one of them is the report [7] stemming from the March 29, 2007, workshop organized by Nanoforum,[a] in which the commercialization of nanotechnology and its key challenges have been addressed. According to the report, more than 4 billion euros have been devoted to nanotech globally in 2005, and three areas of concern for Europe were identified, as follows [8]:

- A very low proportion (only 3.5%) of the global nanotechnology venture capital was invested in Europe (in 2006).

[a]Nanoforum is a European initiative aiming at linking nanotechnology-related activities within the European Union. More information is available at http://www.nanoforum.org/.

- In terms of publications, the European Union (EU) and the United States are pretty much equal, but way few patents have been granted to EU-based applicants.
- The level of industrial investment in nanotechnology from the EU is almost half of that offered by the United States and Japan. According to the report, the EU industry invested 0.9 billion euros in 2005, as opposed to 1.8 billions of Japan and the United States.

In the report it is also highlighted that in terms of patent production, the EU lags behind the United States, especially because of the kind of research that is performed, which falls in the realm of basic or pure research. At the workshop, and in the report then, it has also reported that the first-to-file system adopted in the EU is not that conducive to patent production, and researchers, especially in Europe, still face the patent dilemma, questioning themselves whether it would be better to publish their work first. The report concludes [9] that the best way to favor nanotechnology private funding and commercialization would be to move forward toward a more problem-centric approach to attract further interest from industry.

In the United States, another recent report [10] adopting the Delphi method[a] revealed that nanotechnology should learn from its bigger brother, that is, microelectromechanical systems (MEMS). The major findings of this research showed that nanotechnologies should follow these basic rules to grow smoothly:

- "not to create technology for mere technology sake;
- understand market's unfulfilled needs;
- understand competitive offerings and create a different product;
- not overstate the ability of nanotechnology to solve problems, miraculously;
- properly promote novel products" [11].

---

[a]The Delphi method is an interactive forecasting method that relies on a panel of independent experts. They are carefully selected and they answer questionnaires in two or more rounds. After each round, they are given an anonymous summary of their forecasts from the previous round so that they can provide revisions to their earlier work.

Another report [12] of 2007 commissioned by the US Department of Commerce addressed the barriers of nanotechnology commercialization. The categories of topics distributed to the participants were:

- "Capital issues/market readiness;
- Regulation/environmental, health and safety issues;
- Public attitudes and perceptions;
- Other issues (workforce, standards, manufacturing, infrastructure)" [13].

The major findings stemming from the research are summarized here below [14]:

- Capital issues/market readiness: It is difficult to think that venture capitalists (VCs) would be interested in ventures that take more than three years to pay out. There is a need to merge public and private funding, which are apparently not working in an organic manner. There is a gap between the value put by research on their results and the real business opportunities envisioned by VCs.
- Intellectual property: Nanotechnology is a platform technology and therefore is pretty different from others. The intellectual property (IP) licensed from universities may not be sufficient to operate a business, but it is vital though. Prosecuting and defending patents may be too much of a burden for a small company.
- Economic development and commercialization: There is a need for a national initiative to build the nanotechnology platform. Federal investments should be more planned and involve universities.
- Workforce development and education: It is fundamental to have skilled people to achieve the envisioned results. Some talented researchers are not US citizen/residents and therefore have no access to federal labs, and this is a problem. It is key to provide business management training to nanotech entrepreneurs [15].
- Occupational health: Health and safety issues should be dealt with very seriously and right away to prevent fear

of nanotechnology. Specific laws and regulations may be required.

- Public policy and health: Some people from the public and the press are afraid of nanotechnology because they do not know it. The use of nanotechnology in products should be labeled so that consumers can make informed choices. People seek involvement from the government in the decision-making process. New laws and regulations are necessary to guarantee to the public that there is no harm in products incorporating nanotechnologies.
- Nanotechnology standards: Nanotechnology standards must be established. Start-ups need assistance from standard-setting bodies. Standards are necessary for research, production, and disposal of nanoparticles.
- Risk management: An integrated system in the United States is necessary to address all the health and safety issues of federal agencies involved in nanotechnology.
- Environment: There is a need to further investigate the potential consequences related to the use of nanoparticles already used in industry.

As to the impact of nanotechnology on companies, it is worth mentioning that the Organisation for Economic Co-operation and Development (OECD) has recently issued a publication [16] with insights from several case studies. The main findings, which allegedly differ from previous studies on the subject, may be summarized as follows [17]:

- Nanotechnology is an enabler of both new products and processes, and new services as well.
- Nanotechnology may be employed to better existing products and services.
- Small companies are more prone to dealing with universities to develop their nanotech pipelines as opposed to large ones.
- Nanotechnology can help address broader socioeconomic concerns like clean water, accessible health care, etc.
- Large companies have been able to assimilate nanotechnology research pipelines based on their existing capabilities in

terms of R&D and production, and this runs counter to the traditional company dynamics and technology life cycles.

- Nanotechnology is highly interdisciplinary and requires many different skills; therefore, small companies with limited trained personnel may face some problems in fully developing new products and processes.
- There are several challenges in funding as far as R&D activities are concerned, especially due to the long scalability process [18].
- There are serious challenges for small companies in presenting new technologies to potential customers, especially when different applications in different industries are pursued.
- IP might pose serious issues in the future due to the broad claims of the first patents, which may create barriers to entry for new companies.
- Concerns about the environment, safety, and health issues complicate massively the business environment.
- Commercialization raises great challenges [19]; poor scalability of nanotech is probably one of the most serious challenges, together with the lack of well-trained personnel.

Another major issue about the further development and commercialization of nanotechnologies concerns the role of venture capital [20]. The great potential of nanotechnology in the next initial public offering (IPO) wave has been already envisioned. Garrett [21], for example, analyzes investment trends in nanotechnology and argues that the companies that are most likely to see an IPO [22] are those with commercially viable products and a strong IP, especially strong relationships with large corporations. Chin [23] provides an insightful set of instructions for private equity funds in order to avoid the common pitfalls when investing in nanotech ventures. He also notes that US investments have shorter exit times than European ones. US firms [24] receive larger funding and US VCs are able to manage quicker exits. The reason for this seems to be related to the fact that European VCs are more specialized in closing deals rather than being managers and this should be one of the reasons why the US style appears to be more successful. One of the largest studies on nanotechnology VC-backed companies is the one prepared by Munari and Toschi [25]. They analyzed in their work a sample of 332 VC-backed companies in the 1985–2006 period. These here below are their major findings:

- Financing: On average, companies in their sample received 5.01 million dollars in their first financing round [26].
- Patents: At the time of the first investment, the companies had a maximum of 15 patents (in general) and 7 patents (in the nanotech field), and the mean was 0.84 patents (in general) and 0.28 patents (in the nanotech field). Just 28% had patents in other fields and only 10% owned patents in nanotech. The study also shows that IP is not that fundamental in the first round, but it is definitely important if in the portfolio there are nanotech-specific patents [27].
- VC competencies: The study shows that VCs with specific skills in nanotech have more opportunities to appreciate the value of the idea and patents and therefore more prone to invest [28].

As to the interest attached to nanotechnology valorization and exploitation, though, it is sufficient to say that in Europe, as early as in 2004, the European Commission addressed the importance of nanotechnology in the field of patenting in the communication "Towards a European Strategy for Nanotechnology" [29], in which, among other things, it is stated that the "the management of IPR can be challenging in a field such as nanotechnology where interdisciplinarity brings together researchers and industrialists with different cultures and attitudes" [30].

Indeed, the European Parliament shared the Commission's view and in a 2006 report stated that [31] "protection of intellectual property rights (IPR) in the field of N&N is essential for innovation, both in terms of attracting initial investment and ensuring future revenue; calls on the Commission to develop standards for the protection of IPR and models for licensing agreements" [32].

Thus, the theme of the commercialization of nanotechnologies has been definitely dealt with and still requires a lot of attention if all the identified barriers have to be overcome. In the next section, this study offers a solution for the most common practice of commercialization and therefore the one involving technology transfer from universities and research centers to companies that are then supposed to further develop and commercialize to the public the technologies when they ultimately become proven industrial processes and/or finished products.

9.2 Cost, Price, and Value: Patent Valuation vs. Patent Evaluation

Depending on who you talk to, Facebook is valued at more than $50 billion—maybe even as much as $65 billion. Forbes puts the social networking site's market value higher than Lockheed Martin, Boeing, Target, Sony, Nike and the major automakers. But CNN spoke to some observers who aren't so bullish. "Facebook's not worth $50 billion. I mean, it's just not," according to Douglas Rushkoff, an author and respected teacher on new media. "What people think is that Facebook in the future might be worth more than $50 billion, but for Facebook to be worth more than $50 billion it would have to become a permanent fixture."

—How Much Is Facebook Really Worth?
(CNN Tech, March 08, 2011)[a] [33]

The task of commercializing nanotechnologies is particularly difficult when universities and public research organizations (PROs) are called upon to assign a value to the results stemming from public research in the commercialization phase. In fact, a sale or a license of a technology for a price that is below its potential market value would constitute damage for the national treasury. Since universities and PROs play a dominant role in the commercialization of nanotechnology, as a final step of its internal development, the importance of finding a way to assign a monetary value that could reflect all the benefits and risks of a certain technology becomes

[a] Interestingly enough, Facebook's IPO, which occurred in May 2012, saw the company reaching a market cap of 115 US billion dollars.

critical. Thus, this chapter analyzes the most commonly accepted technology valuation techniques in the literature and currently employed by professionals. It demonstrates how these methods may not be always suitable for nanotechnology, by explaining why these innovations deserve a different approach to value them that could merge quantitative and qualitative assessments at the same time by factoring in different variables that are peculiar to nanotech inventions.

The relationship among nanotechnology, patents, and their value entails the comprehension of several different factors [34, 35]. Due to the fact that nanotechnology research is always trying to push the existing boundaries of science and requires diverse skills, public research institutions, where these skills are present, undisputedly dominate the sector of nanotechnology innovations. However, this implies that universities and PROs are in strong need of monetizing these innovations to replenish their budgets. Therefore, universities and other research institutions have originally seen the patenting of nanotechnology innovations as a new way to monetize their intangible assets after the hype of the biotechnology era. The result is that now there are hundreds of technologies waiting to be licensed or assigned.

The most important thing that must be emphasized at this stage is the crucial difference between three different and distinct concepts: cost, price, and value. The cost is the amount of money necessary to produce a product or perform a process. The price is the amount of money necessary to purchase a product or see a process performed and/or delivered. The price of a product can vary. Take the example of a can of Coke; in a supermarket the price may be 1 euro, in a bar it may be 2 euros, and 3 or more euros in a fancy restaurant. Finally, the value [36] of a product is the price that a person or a company is willing to pay in certain circumstances. So, a can of Coke in the supermarket might be valued at 1 euro because of its abundance, but the same can of Coke would be probably worth 1000 euros to a thirsty traveler lost in the desert. A novel invention may be worth millions of dollars if it is indispensable to a certain market, but it might be valued at 1 euro if there is no market for it or if the technology is already obsolete.

Thus, moving forward, by patent valuation we mean the process of attaching a value, that is, a figure, to a technology. It is therefore a quantitative method. When looking at the kind of technology covered by a patent and at the financing and regulatory hurdles that the technology may encounter before market entry, we are making a qualitative assessment. Such an approach is not related to a figure but to careful considerations drawn after an analysis of the different sectors concerned. Both valuation and evaluation are arbitrary since we make projections of the potential generated income of the technology or we envision what kind of problems can or cannot be solved in its future employment. It is also worth mentioning that when attaching a value to a patent, we necessarily have to consider the potential cost, price, and value of a technology. However, this is not a valid assumption when we perform an evaluation. As mentioned before, in this latter case the assessment is purely qualitative and therefore there is no figure involved. It is true that an evaluation might change the cost, price, and value of the technology, but until now this has not been assessed in great detail because of the generally accepted dichotomy between valuation and evaluation.

Andriessen [37] proposes another triadic approach introducing the concepts of value, valuation, and (knowledge) valorization. The most interesting is the last, which he defines [35, 38] as the utilization and transfer of knowledge to the commercial sector for economic benefit. He conceptualizes knowledge as a thing that might be created, stored, and subject to transactions, and that is why it is possible to talk about value and valuation.

9.2.1 *Patent Valuation*

Valuing a business to obtain funding or receive further investment is a pretty common operation nowadays [39]. Within the realm of business valuation, then, there might be a core technology that drives all the business; this is very common in high-tech start-ups, especially academic ones. In today's professional practice several methods are employed to value a patent [40]. Flignor and Orozco [41] argue that any valuation exercise can be seen as a pyramid in which the foundation blocks are the purpose (what is the valuation for?), the description (what is the asset?), the premise (what is the asset's use?), and the standard (who is the potential buyer?). The valuation purpose may differ greatly. In fact, a valuation maybe needed for financial reporting, a transaction strategy, during merger and acquisition (M&A) operations, litigation, bankruptcy, etc. The valuation description is fundamental as any different type of intangible asset can lead to a different commercialization strategy. For example, a copyrighted work with stable revenues might well be subject to securitization, as it has already happened in the past.[a] The valuation premise regards the use of the asset that will be made in the future, which may be different from the past, especially in the case of bankruptcy. The valuation standard refers to the definition attached to the valuation purpose. The most common standard is the *fair market value*, that is, the price a willing buyer and a willing seller would adopt to complete a transaction. The authors move on by presenting the most common approaches in valuing intangible assets by presenting the following lists:

Traditional methods [42]:

- Transaction: This is based on the so-called fair market value, that is, a price that has been agreed upon for similar transactions.
- Income approach: This is based on the potential cash flows that an asset can generate.
- Replacement cost: This is based on the cost that should be sustained to recreate a fungible asset like the one covered by the legal protection.

Among the newer methods [43]:

- Monte Carlo simulations: These are computational algorithms often used to allow the creation of stochastic or probabilistic financial models.
- Real options valuation [44]: This is a valuation method that, contrary to the discounted cash flow analysis, assumes that the management of a company or the owner of a patent

[a]See the history of David Bowie's bonds, probably the first case of IP securitization, available at http://en.wikipedia.org/wiki/Bowie_Bonds.

is active and different decisions might be made during the life of the asset that can influence its value. In practice, this method takes into consideration the risks as opposed to the income approach.

- Binomial pricing model: This method is carried out by means of using a tree-shaped analysis for a number of time steps between the valuation and expiration dates. Each node in the tree represents a possible price at a given point in time.

Potter [45] illustrates two other valuation methods [46], the so-called technology factor method and the technology risk/reward method. The first takes into consideration the apparent shortcomings of the three major valuation methods as it measures directly the role of the technology within the business being valued; therefore, the final value is not lumped together with the rest of the business's assets. The second uses the value of comparable businesses and then subtracts the amount of money needed to re-create the technology at stake.

Kamiyama et al. [47] divided the spectrum of valuation methods into two macro areas, qualitative and quantitative. The former provides a rating of an asset on the basis of factors such as the strength of the patent and its breadth. The latter gives a monetary appraisal of the asset. They further explain that quantitative techniques are further divided into the cost approach, the market approach, and the income approach (Table 9.1). In their work there is an interesting comparison with all the advantages and disadvantages related to each method [48]. Some of them are worth mentioning as they specifically relate to patent valuation.

9.2.1.1 Cost approach

The cost approach is straightforward. In this case the value of the patent is strictly related to the amount of money spent to conceive the technology. According to this method there are two subsets of rules. Some say that the amount of money spent is what should be taken into consideration, which means the historical costs occurred. Others adopt a different view, arguing that the costs to be

Table 9.1 Advantages and disadvantages of valuation approaches

	Cost approach	Income approach	Market approach
Advantages	Objective and consistent	Extra value as based on future earnings	Practical and easy approach
Disadvantages	No correlation between the expenditure and the asset's value	Assumptions pretty limitative as the management is passive Relevant information not always available	Details of patent transactions very infrequently known

considered should be accrued with the relevant interest until the day the negotiation takes place. A third option might be to assess how expensive it would be to conceive the technology today. It may seem irrelevant, but as the following scenarios will illustrate the numbers may vary greatly. Let us assume that company A developed a technology in 1999 by investing a million dollars, and company B, to reproduce the same technology in 2010, would need to spend $500,000 dollars. If we want to consider inflation and update the price of goods and services, the amount would rise to $1,312,791.12.[a] So, it becomes clear that the subapproach the parties choose to adopt, when valuing their patents, is not that trivial. In our case company B could pay for the same technology between $500,000 and $1,312,791.12.

9.2.1.2 Market approach

In theory, the market approach is the easiest approach. It assumes that the value of a piece of IP should be equal to the value it would yield in the marketplace. So, what is fundamental is to take a look at similar transactions that occurred in the market. This task is difficult in practice, since this type of information is not easily accessible. Negotiations about IP are oftentimes kept highly confidential and only rarely are there leaks concerning the details of the transactions.

[a]Amount obtained using the inflation calculator from http://data.bls.gov/cgi-bin/cpicalc.pl (last accessed on January 18, 2011). The consumer price index (CPI) inflation calculator uses the average CPI for a given calendar year.

In fact, in most cases even the disputes concerning these assets are settled through arbitration, and therefore, the details of the cases are kept secret.

9.2.1.3 Income approach [49]

The income approach is divided into different sub-approaches, royalty relief and discounted cash flow [50], to find the present value of a technology. Both theories are related to the potential income the user of the technology might accumulate over a certain period of time. Relief from royalty provides that if a company loses ownership of a particular intangible asset, it has to pay a royalty to license it from someone else. Under this method, the value of the patent is the capitalized value of the royalties the company does not have to pay when it owns the patent. However, the amount of money to transfer should be calculated at present, and the amount of royalties given in a certain period must be discounted using a predetermined rate to find the present value of the technology. Thus, first the sum of the projected cash flows must be identified to then be discounted using an interest rate. This is usually the rate linked to treasury bonds or other more or less stable indicators. The discounted cash flow method is based on the same principle and can be used for both discrete cash flows and multiple cash flows. It is worth emphasizing that this method is probably more useful for the type of technology that is an inherent and necessary part of the production process, in addition to contributing to the quality of a product and by extension boosting its sales.

Some considerations about nanotechnology must be mentioned with regard to the delineated approaches. Nanotechnology does not just bring novel features into our homes and our daily lives; it is strongly intertwined with different regulatory aspects that still need to be resolved. For example, it is still not clear whether products incorporating nanomaterials should comply with certain specific rules in the United States and Europe. Sure, numerous regulatory agencies have circulars and internal directives, stating that they have round tables and working groups about nanotechnology. However, in truth, there is currently a lot of uncertainty surrounding the use of nanomaterials in everyday products, both for consumers and for producers. There are not enough studies addressing the potential toxicity of the products and the few that do exist often come to conflicting conclusions. So, when attaching a value to a product incorporating nanomaterials more attention should be paid to the underlying technology. In fact, when using the income approach, which is probably the most common, the projections of future cash flows are directly dependent on the projected sales. These might be heavily hindered by, for example, more or less stringent regulatory provisions in the future.

Also, when thinking about future projections, we must rely on assumptions that may or may not be true. This is a common principle for all products or services, but nanotechnology is different. In this world made of invisible components, unexpected events can occur that may adversely affect the life of a product or service. Let us take a quick example. It might be that a novel nanotechnology-based paint for cars that guarantees antiscratch properties would actually result in a product that is highly inclined to corrosion after a couple of years. Nanotechnology is in its infancy, so there might not be enough time or knowledge to assess the side effects of the novel products being marketed right now. Having said that, let us think for a minute about the potential consequences in this case. The paint producer or the car manufacturer could be obliged to repaint the cars after the corrosion if it occurred during the warranty period. This is a real-world scenario that might happen with nanotechnology-based innovations, since the unexpected properties displayed by these products may well have unexpected consequences. An unexpected event like the one just mentioned would not only temporarily stop the sale of the cars employing the novel paint; it would also tarnish the company's image, which can be even worse than a potential drop in its sales. This illustrates that the common income approach might not be the right way to go for nanotechnology-based innovations due to their "volatile" nature at this point.

Coleman [51], in a recent work, illustrates a novel hybrid valuation method that takes into consideration the peculiarities of nanotechnology. He assumes [52] that nanotech companies are poor in cash flow, they have few hard assets, and the market is still very small. Therefore, he introduced additional inputs to take into considerations. Some of these inputs include company asset-specific research, industry and market research, financial analysis, corroboration of valuation results, and capability benchmarks. Then, since nanotechnology is creating a totally new market, he provides an impact-scoring system based on four variables: business, government, society, and education.

Wartburg and Teichert [53] tried to provide a solution for the valuation of nanotechnology-related assets. They assume that nanotech patents will serve as an important asset for future securitization operations. They also argue that the nanotech patent landscape is overcrowded, especially as to the patents covering alternative structures at the nanoscale. Additionally, they think that considering the whole value chain, those interested in the commercialization of nanotech products will have to take into consideration quite complex patent strategies allowing patent pooling, IP insurance, and cross-licensing [54]. Lastly, they conclude by advocating the cooperation among the stakeholders along the entire value chain to create paths that might be viable and economically profitable [53].

9.2.2 *Patent Evaluation*

The term "evaluation" implies something more than a figure attached to a certain object. In fact, when we evaluate something we consider different variables, and the kind of output is qualitative and not purely quantitative as in the case of valuation.[a] We might think that when evaluating a technology, the most important variables to look at are the market, regulatory and legal issues, the technology's future, and financing opportunities.

9.2.2.1 Market

The market is the most important factor in a business plan and it is the most important variable when assessing the potential of a technology. Therefore, a promising technology will be even more so if the barriers to entry in the market are low due to little or no

[a]For another general overview of patent evaluation, see the EPO webpage, Patent Portfolio Management and Patent Evaluation, available at http://www.epo.org/patents/patent-information/business/valuation/faq.html.

competition. On the other hand, the competitive advantage of being the first one in a market could turn into a disadvantage. After all, the forerunner is clearing the way for its future competitors who had the opportunity to learn from its mistakes. Another important factor to consider is whether the market is ready for a new product, since its success cannot be taken for granted.

9.2.2.2 Regulatory and legal issues

When evaluating a technology it is crucial to take all the regulatory and legal issues into consideration. In fact, a technology might well be groundbreaking and have the potential to improve the consumer's quality of life, but if there are regulatory and legal barriers that do not or only partly allow its commercialization, the technology's value will be much lower than expected. For example, there are a lot of technologies nowadays based on stem cells, but at the same time many jurisdictions do not allow their experimentation. So, in these places the technology is theoretically worth nothing. As a consequence, environmental and safety-related aspects of a certain technology will probably be the major hurdles to be overcome in terms of legal and regulatory barriers.

9.2.2.3 Technology

When evaluating a technology it is equally important to analyze its long-term potential, since there is always a risk it will become obsolete or not used at all in the close or distant future [55]. Keeping this in mind, it is key to create different future scenarios to assess the life expectancy of the technology and its utilization. Just to give an example, it is foreseen that by 2020 every household in the United States will have a 3D printing machine. The feedstock for these printers today consists primarily of plastics. At the same time plastics are destined to give way to bioplastics in the near future. From this assumption we can, therefore, infer that a technology related to bioplastic materials for 3D printers will probably have a bright future.

9.2.2.4 Financing

Financing issues are crucial when the technology and its owner are looking for potential investments. Say that the business plan of a future company relies on a strong patent/technology that is easily funded due to its potential future success. It will be easier for the would-be entrepreneur to find business angels or VC firms that believe in the investment, and they will probably be more prone to invest in the business idea.

In light of the above, we can easily comprehend how evaluation is a broader concept that takes into consideration more variables than the mere valuation of a product or process. In general terms, evaluation is a complex analysis that is meant to be like a strengths, weaknesses, opportunities, and threats (SWOT) analysis, which ponders the endogenous and exogenous components of the object to be examined. Among all technology fields, nanotechnology is probably the most suitable for evaluation techniques. In fact, nanotechnology opens the door to endless issues relating to the evaluation of technologies. Indeed, there is so much uncertainty on the regulatory issues or unexpected behaviors surrounding it that all the different variables might be affected. For instance, if a new nanomaterial was able to create a whole new market from scratch, the regulatory aspects around its use will be crucial. A novel drug delivery method to cure cancer, which employs nanomaterials, for example, might be a breakthrough. However, if the Food and Drug Administration (FDA) or the European Medicines Agency (EMEA) would find it not suitable for human safety due to its confirmed or potential toxicity, it might lead the business in question to bankruptcy if the commercialization is stopped or heavily delayed for safety concerns. To sum up, the evaluation of nanotechnology-related inventions is probably the most difficult to determine as the future of these innovations is still uncertain in terms of latitude, regulations, drawbacks, and employment.

Therefore, due to the very nature of nanotechnology innovations, it may be safe to say that the only way to provide a reliable judgment about the economic potential of these products is to avoid the use of valuation techniques and rely on evaluation as an adherent method to gauge potential real-world scenarios. Unfortunately, evaluation is a qualitative analysis, and therefore, there is no figure attached to it, which is why a novel method that could provide a merger between a qualitative analysis and a valuation with adjustable variables is needed. This novel method is illustrated in detail in the next section.

9.3 Introducing a Novel Approach: "Present Value After Evaluation" [56]

In the quest to comprehend what might be a reasonable value to attach to a technology or patent, a novel evaluation method and associated tool (i.e., a custom worksheet) called **Present Value After Evaluation** [57] (PVAE) is being presented. This novel approach takes the present value of a patent or technology into consideration by using the discounted cash flow method and then adds the following additional variables using a scale from 1 to 5:

- Patent relevance (considers the relevance of the patent in the final product or process)
- Patent coverage (considers the strength of the patent according to the claims and existing or potential litigation)
- Technology (considers the technology's future scenarios)
- Financing (considers the attractiveness for investors)
- Regulation (considers regulatory and legal barriers)
- Market (assesses the potential success of the product or process)

The model contains other five variables that should be made when using the PVAE tool, namely:

- Growth rate (three variables)
- Discount rate
- Royalty rate
- Tax rate
- Cost of goods sold (COG) and overheads

To have a more detailed idea of how the PVAE method works, the following table (Table 9.2) explains in detail the nature of the variables and the formulas behind them.

Table 9.2 Explanation of the PVAE tool's variables

Variable	Explanation
Patent relevance	It measures the importance of the patent/technology in the product/process (1 = poor; 5 = excellent). Every point assigned to the technology increases or decreases the PVAE by 20%. It is assumed that the technology has a minimum relevance of 20% in the product/process in which it is incorporated. This variable can be applied also to nonpatented technologies. In this case the value to be assigned to the technology should be 1, assuming that the technology, even if not patented, may still get protected through patents over its future improvements.
Patent coverage	It measures the quality in terms of capability to resist in a patent opposition or infringement lawsuit of the patent in the product/process (1 = poor; 5 = excellent). Every point assigned to the technology increases or decreases the PVAE by 20%. It is assumed that the patent guarantees a minimum coverage of 20% for the product/process that is patented. This variable can also take into consideration the different scenarios of the applications being prosecuted before different patent offices. This variable can be applied also to nonpatented technologies. In this case the value to be assigned to the technology should be 1, assuming that the technology, even if not patented, is not infringing on someone else's IP.
Growth rate	It indicates the projected increase in sales. It offers three settings, which work at the same time during the life of the technology/patent. This way the user can think of different scenarios over the life of the technology/patent.
Discount rate	It is necessary to calculate the present value. The discount rate is agreed upon by the parties considering other transactions and the risk associated with the technology. The higher the risk, the higher the discount rate.
Royalty rate	It should be consistent with industry standards. The royalty rate, according to the relevant industry, can easily range between 0.1% and 60% of the sale price; therefore, this is a critical variable that must be agreed on by the parties.
Tax rate	It indicates the potential tax rate to be applied. It varies according to the jurisdiction in which the taxes are to be paid.
COG, etc.	It includes all costs, including overheads and expenses, related to the production of the product or performance of the process in question.
Technology	It measures the potential success of the patent/technology, including future scenarios (1 = poor; 5 = excellent). Every point assigned to the technology variable increases or decreases the present value of the cash flows, considering the relevance of the patent and patent coverage by 5%, and contributes to provide the PVAE. It is assumed that the technology in question has a minimum potential future equal to 5% of the PVAE.

Financing	It measures the attractiveness of the patent/technology for potential investors (1 = poor; 5 = excellent). Every point assigned to the financing variable increases or decreases the present value of the cash flows, considering the relevance of the patent and patent coverage by 5%, and contributes to provide the PVAE. It is assumed that the technology in question has a minimum potential in terms of an investment target equal to 5% of the PVAE. If there is no interest in approaching potential investors, the value should be 5.
Regulation	It measures the regulatory and normative barriers for the sale of the product or performance of the process (1 = poor; 5 = excellent). Every point assigned to the regulation variable increases or decreases the present value of the cash flows, considering the relevance of the patent and patent coverage by 5%, and contributes to provide the PVAE. It is assumed that the technology in question has a minimum potential in terms of marketability (from a regulatory perspective) equal to 5% of the PVAE. Regulatory provisions are related to each jurisdiction or region (e.g., the EU); therefore, in the case of a patent family, the user can opt for an average value after careful consideration of all the jurisdictions concerned or he or she may opt for an individual valuation for each single country or region. In this case, the final PVAE will probably be an average of all the PVAEs related to each single patent of the patent family.
Market	It measures the market potential (in terms of sales) of the technology (1 = poor; 5 = excellent). Every point assigned to the financing variable increases or decreases the present value of the cash flows, considering the relevance of the patent and patent coverage by 5%, and contributes to provide the PVAE. It is assumed that the technology in question has a minimum potential in terms of marketability (from an economic perspective) equal to 5% of the PVAE. At times, market data is related to specific countries or regions (e.g., the EU); therefore, in the case of a patent family, the user can opt for an average value after careful consideration of all the countries concerned or he or she may opt for an individual valuation for each single country or region. In this case, the final PVAE will probably be an average of all the PVAE values related to each country or region. In measuring this variable, it may be useful to use Porter's five force analysis [58], as shown in Table 9.3.
Total PV of cash flows, considering the relevance of the patent and patent coverage	It measures the present value of the technology/patent, considering the relevance of the patent and patent coverage. It may be equal to the present value of cash flows as long as the relevance of the patent and patent coverage has the maximum score (i.e., 5).

(*Contd.*)

Table 9.2 (*Contd.*)

Variable	Explanation
PVAE	It measures the present value, considering all the other variables explained above. Therefore, it represents at the same time a quantitative assessment (expressed in $ or other currency) but after weighing qualitative variables related to the life and use of the technology/patent.
Initial cash flow	It is the projected amount of sales for the first year of operation by using the technology/patent.

Table 9.3 Results of Porter's five force analysis

Porter's five forces analysis	Rating 1.0–5.0
The threat of the entry of new competitors	2.9
The intensity of competitive rivalry	3.6
The threat of substitute products or services	3.3
The bargaining power of customers (buyers)	3.3
The bargaining power of suppliers	2.8
Average	**3.1**

Ideally, to find the proper score for the market variable it may be useful for the user to run Porter's five force analysis, which considers the likelihood of a new business to be successful in a given industry; the higher the score, the lower the likelihood of success. Below, in Table 9.3, the five pillars of the analysis are enunciated in detail:

- The threat of the entry of new competitors
- The intensity of competitive rivalry
- The threat of substitute products or services
- The bargaining power of customers (buyers)
- The bargaining power of suppliers

By arbitrarily assigning a value from 1 to 5 (to stick to the ranges of the market variable in the PVAE worksheet), it is possible to have a final average value that can be used for the market variable by rounding it down or up according the value of the decimal. Every industry force, in the PVAE worksheet, has five additional subcategories that can be rated, which is why the values in the right column might have decimals.

In the PVAE worksheet, we must start with an initial projection of the first cash flow. When populating the sheet, we have to make several assumptions concerning the discount rate, growth rate of the projected sales, royalty rate, tax rate, and percentage of COG and other expenses. All the other variables are modifiable using a 1 to 5 scale in which 1 is "poor" and 5 is "excellent," as it can be easily inferred from Table 9.4 below. Patent coverage and patent relevance are directly linked to the total present value of cash flows as opposed to the other variables that occupy a second layer of importance and provide an average output. In the example shown in the next table (Table 9.4), we assumed that the initial cash flow is equal to $100,000 and that all the variables are set on "excellent" (meaning that the technology is necessary to make the product or implement the process), that the claims of the patent have a wide and protectable breadth, that the technology is promising, and that there are no regulatory and market barriers. So, we are dealing with an excellent technology in this case. Considering a life expectancy of 15 years, the total of cash flows is $2,867,778 and the total present value is $2,192,010. In this case, the PVAE after tax, COG, and other expenses is equal to $1,150,805. The sheet also provides the amount of royalties collected over time and their present value, which in this case is $107,226. Thus, in the projected scenario, the present value of the royalties is almost one-tenth of the PVAE after tax, COG, and

Table 9.4 Table showing the variables chosen for the given technology/patent

Assumptions			
Patent relevance 1–5:	**5**		
Initial cash flow:	$100,000		
Patent coverage 1–5:	**4**		
Years:	1–5	6–10	11–15
Growth rate:	**30%**	**20%**	**20%**
Technology 1–5	Financing 1–5	Regulation 1–5	Market 1–5
5	**5**	**5**	**1**
Discount rate:	**40%**	**Tax rate:**	**35%**
Royalty rate:	**10%**	**COG, etc.**	**50%**

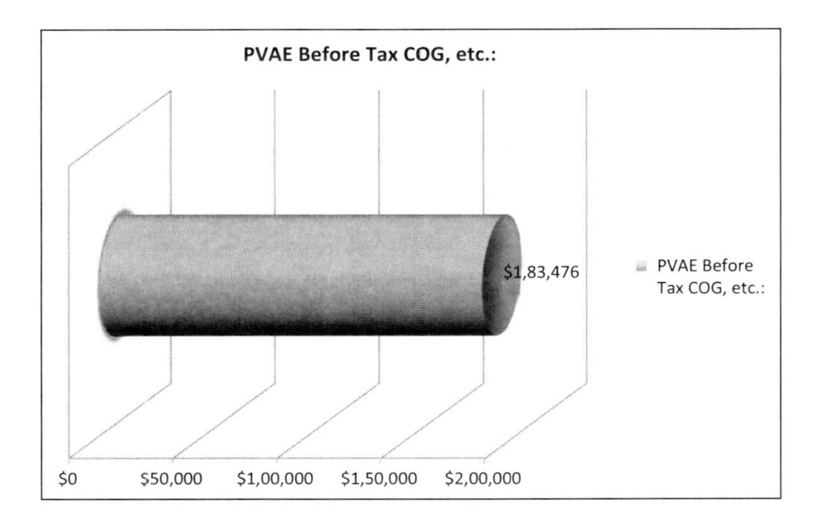

Figure 9.1 PVAE chart before tax, COG, etc.

other expenses. This is an amazing outcome because it proves that for a very promising technology, the assignee would earn 10 times as much than what was paid to the original patentee if the price of the transaction is equal to the present value of the prospected royalties. In the chart below (Fig. 9.1), automatically produced by the PVAE worksheet, the user can have a graphical overview of the values assigned to the six major variables in this scenario. The selected variables are represented in the PVAE worksheet that follows (Table 9.5).

In the first scenario it is easy to understand from the table above that the technology/patent at stake is quite important for the implementation of the final product/process, very strong from a legal standpoint, very promising in terms of future development, definitely risk free from a regulatory perspective, but destined to enter a very competitive market [59].

The PVAE worksheet also automatically generates useful graphs that allow the user to have a graphical perspective of the different scenarios. In the case at stake, the graphs generated by the worksheet are shown in Figs. 9.2–9.5.

In the second scenario (Table 9.6), we populated our table using a different evaluation, and we graded the technology by attaching a 3

Table 9.5 PVAE worksheet in the first scenario

Present value after evaluation							
Patent relevance 1–5:	5						
Initial cash flow:	$100,000			Technology 1–5	Financing 1–5	Regulation 1–5	Market 1–5
Patent coverage 1–5:	4			5	5	5	1
Years:	1–5	6–10	11–15				
Growth rate:	30%	20%	20%				
	Before tax, COG, etc.	After tax, COG, etc.					
Total of cash flows	$12,741,535	$4,140,999	Discount rate:	40%		Tax rate:	35%
Total PV of cash flows:	$728,080	$236,626	Royalty rate:	10%		COG, etc.	50%
Total PV of cash flows considering the relevance of the patent and patent coverage:	$582,464	$189,301	Partial PVAE	$582,464	$582,464	$582,464	$116,493
Present value of royalties (after tax)	$5,324		PVAE before tax COG, etc.:	$465,771	PVAE after tax, COG, etc.:	$151,441	
Year	Flows	Growth	Present value	Year	Rate	Royalty-tax	Total
1	$130,000	30%	$92,857	1	10%	$8,450	$8,450
2	$169,000	30%	$86,224	2	10%	$10,985	$19,435
3	$219,700	30%	$80,066	3	10%	$14,281	$33,716
4	$285,610	30%	$74,347	4	10%	$18,565	$52,280
5	$371,293	30%	$69,036	5	10%	$24,134	$76,414
6	$445,552	20%	$59,174	6	10%	$28,961	$105,375
7	$534,662	20%	$50,720	7	10%	$34,753	$140,128
8	$641,594	20%	$43,475	8	10%	$41,704	$181,832
9	$769,913	20%	$37,264	9	10%	$50,044	$231,876
10	$923,896	20%	$31,941	10	10%	$60,053	$291,929
11	$1,108,675	20%	$27,378	11	10%	$72,064	$363,993
12	$1,330,410	20%	$23,467	12	10%	$86,477	$450,470
13	$1,596,492	20%	$20,114	13	10%	$103,772	$554,242
14	$1,915,790	20%	$17,241	14	10%	$124,526	$678,768
15	$2,298,948	20%	$14,778	15	10%	$149,432	$828,200

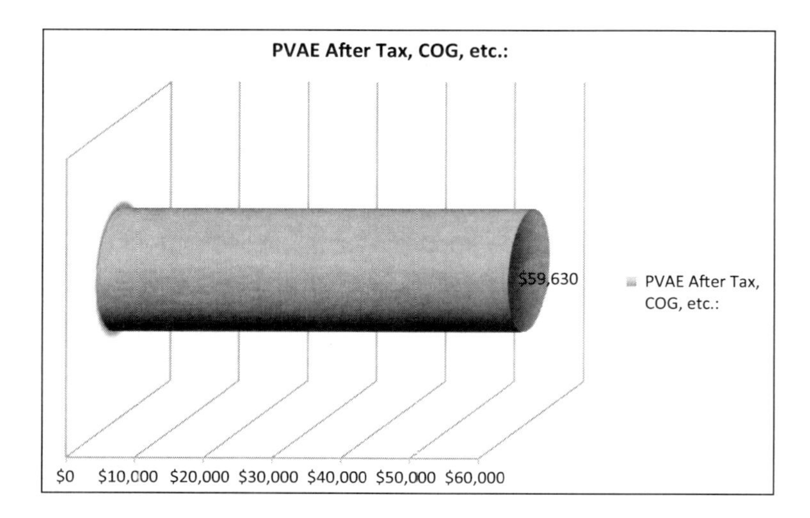

Figure 9.2 PVAE chart after tax, COG, etc.

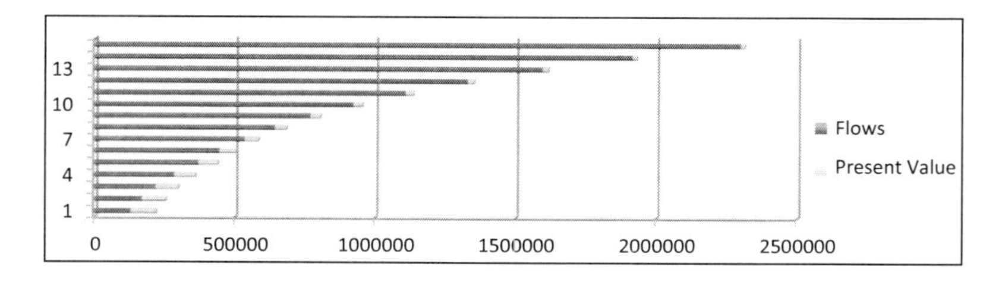

Figure 9.3 Future cash flows and present value.

to all the variables. This means that the technology in question is not that crucial to the production of the product or the implementation of the process. It also implies that the claims of the patent have a medium breadth, that the technology is moderately promising, and that there might be some regulatory and market barriers. Starting with the same initial cash flow for the first year of sales, and considering a life expectancy of 15 years, the total of cash flows is again $2,867,778, and the total present value remains at $2,192,010. However, here the total present value considering the relevance of the patent and patent coverage is $789,123 and the PVAE after tax, COG, and other expenses is equal to $248,574. The sheet also provides the amount of collected royalties and their present value,

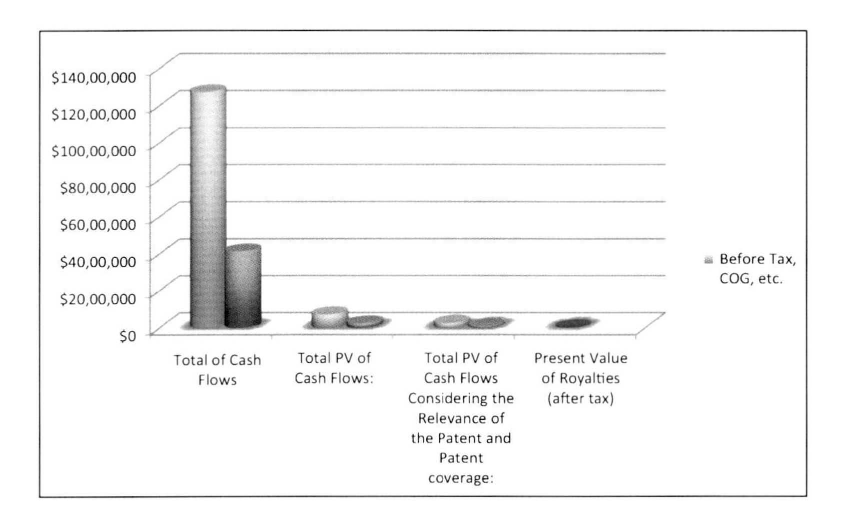

Figure 9.4 Cash flows and royalties.

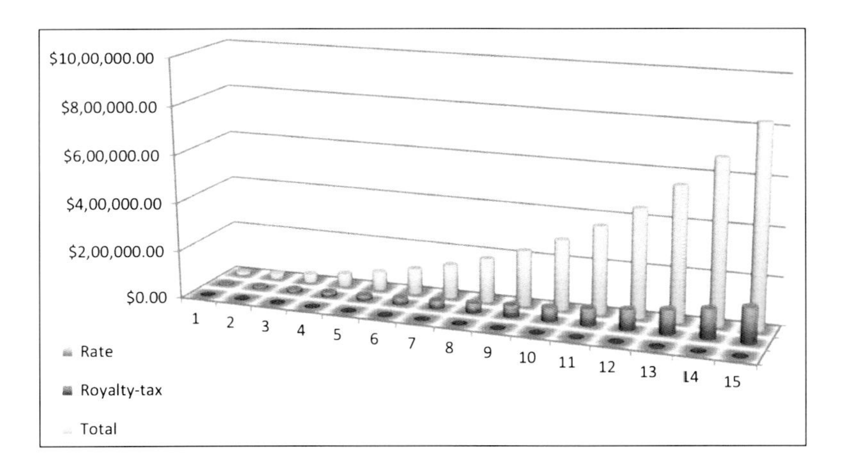

Figure 9.5 Royalty revenues.

which is again $107,226. So, in this second scenario, the present value of the prospected royalties is more than one-third of the PVAE after tax, COG, and other expenses. In the chart below (Fig. 9.6), the user can have a graphical overview of the values assigned to

Table 9.6 Variables chosen for the given technology/patent

Assumptions			
Patent relevance 1–5:	3		
Initial cash flow:	$100,000		
Patent coverage 1–5:	3		
Years:	1–5	6–10	11–15
Growth rate:	**30%**	**20%**	**20%**
Technology 1–5	Financing 1–5	Regulation 1–5	Market 1–5
3	5	3	3
Discount rate:	**40%**	**Tax rate:**	**35%**
Royalty rate:	**10%**	**COG, etc.**	**50%**

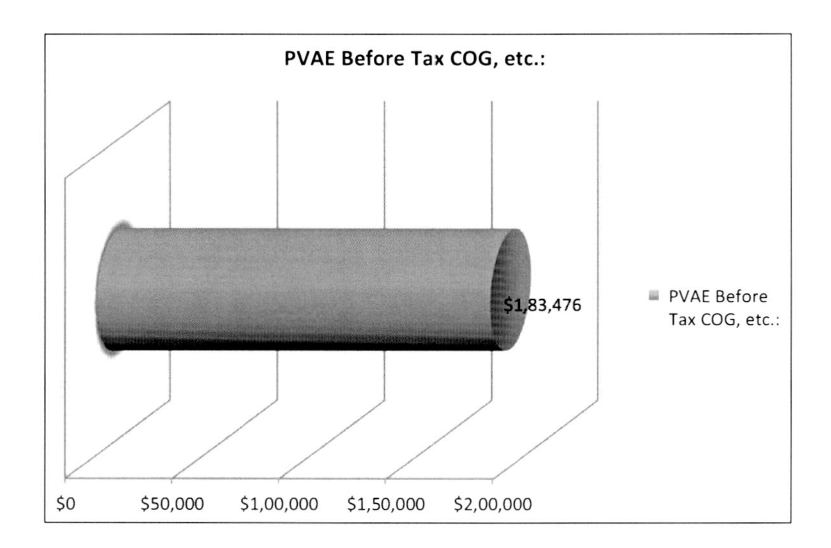

Figure 9.6 PVAE chart before tax, COG, etc.

the six major variables in this scenario. The selected variables are represented in the PVAE worksheet (Table 9.7).

In the second scenario it is easy to understand from the table above that the technology/patent at stake is of average importance for the implementation of the final product/process, possessing an average strength from a legal standpoint, promising in terms of future development, somehow risky from a regulatory perspective, and destined to enter a somewhat competitive market [59].

Table 9.7 PVAE worksheet in the second scenario

Present value after evaluation							
Patent relevance 1–5:	3						
Initial cash flow:	$100,000			Technology 1–5	Financing 1–5	Regulation 1–5	Market 1–5
Patent coverage 1–5:	3			3	5	3	3
Years:	1–5	6–10	11–15				
Growth rate:	30%	20%	20%				
	Before tax, COG, etc.	After tax, COG, etc.					
Total of cash flows	$12,741,535	$4,140,999	Discount rate:	40%		Tax rate:	35%
Total PV of cash flows:	$728,080	$236,626	Royalty rate:	10%		COG, etc.	50%
Total PV of cash flows Considering the relevance of the patent and Patent coverage:	$262,109	$85,185	Partial PVAE	$157,265	$262,109	$157,265	$157,265
Present value of royalties (after tax)	$5,324		PVAE before tax COG, etc.:	$183,476	PVAE after tax, COG, etc.:	$59,630	
Year	Flows	Growth	Present value	Year	Rate	Royalty-tax	Total
1	$130,000	30%	$92,857	1	10%	$8,450	$8,450
2	$169,000	30%	$86,224	2	10%	$10,985	$19,435
3	$219,700	30%	$80,066	3	10%	$14,281	$33,716
4	$285,610	30%	$74,347	4	10%	$18,565	$52,280
5	$371,293	30%	$69,036	5	10%	$24,134	$76,414
6	$445,552	20%	$59,174	6	10%	$28,961	$105,375
7	$534,662	20%	$50,720	7	10%	$34,753	$140,128
8	$641,594	20%	$43,475	8	10%	$41,704	$181,832
9	$769,913	20%	$37,264	9	10%	$50,044	$231,876
10	$923,896	20%	$31,941	10	10%	$60,053	$291,929
11	$1,108,675	20%	$27,378	11	10%	$72,064	$363,993
12	$1,330,410	20%	$23,467	12	10%	$86,477	$450,470
13	$1,596,492	20%	$20,114	13	10%	$103,772	$554,242
14	$1,915,790	20%	$17,241	14	10%	$124,526	$678,768
15	$2,298,948	20%	$14,778	15	10%	$149,432	$828,200

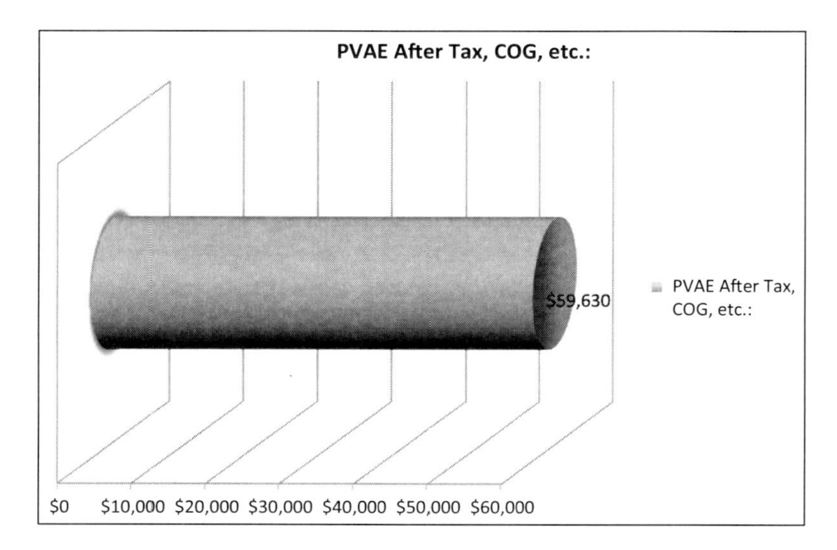

Figure 9.7 PVAE chart after tax, COG, etc.

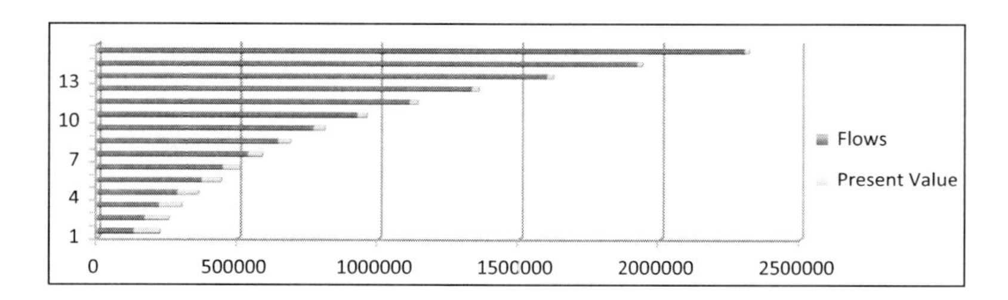

Figure 9.8 Future cash flows and present value.

Once again, the PVAE worksheet also automatically generates useful graphs that allow the user to have a graphical perspective of the different scenarios. In the case at stake, the graphs generated by the worksheet are shown in Figs. 9.7–9.10.

In the third scenario, the table (Table 9.8) is populated with different future variables, and the technology is graded by attaching a 2 to all the variable slots. The technology in question is, therefore, definitely not crucial to the production of the product or the implementation of the process. It also means the claims of the patent have a medium-low breadth, the technology is not that promising,

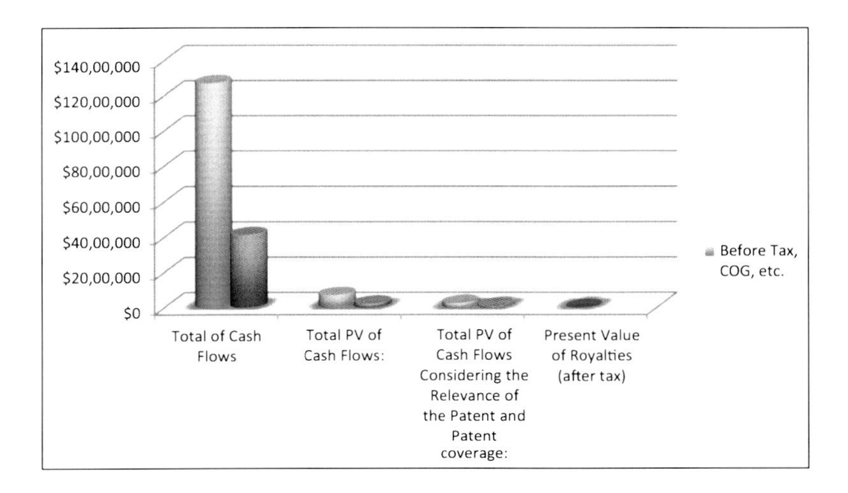

Figure 9.9 Cash flows and royalties.

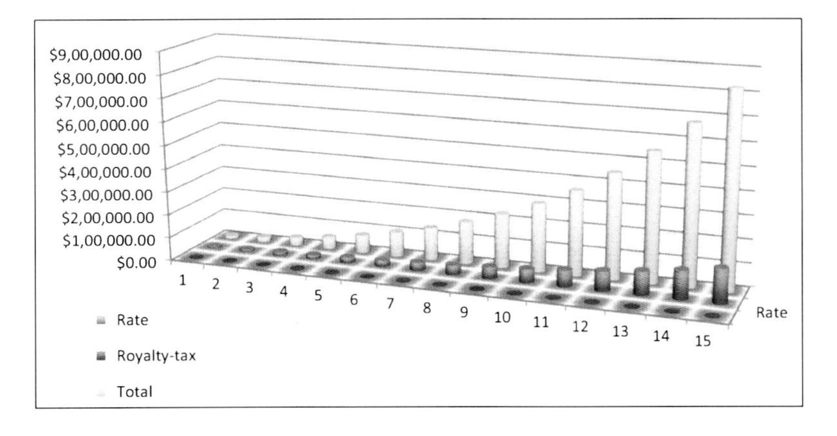

Figure 9.10 Royalty revenues.

and there might be some serious regulatory and market barriers. Starting with the same initial cash flow for the first year of sales, and considering a life expectancy of 15 years, the total cash flows are once again $2,867,778, and the total present value is still $2,192,010. However, here the total present value considering the relevance of the patent and patent coverage is $350,722, and the PVAE after tax, COG, and other expenses is equal to $73,652. The sheet also provides

Table 9.8 Variables chosen for the given technology/patent

Assumptions			
Patent relevance 1–5:	2		
Initial cash flow:	$100,000		
Patent coverage 1–5:	2		
Years:	1–5	6–10	11–15
Growth rate:	30%	20%	20%
Technology 1–5	Financing 1–5	Regulation 1–5	Market 1–5
2	5	2	2
Discount rate:	40%	Tax rate:	35%
Royalty rate:	10%	COG, etc.	50%

the amount of collected royalties and their present value, which is again $107,226 because they are based on the same amount of sales. Thus, in this last scenario, the present value of the foreseen royalties is considerably higher than the PVAE after tax, COG, and other expenses. These three examples clearly show that using the present value of expected royalties to determine the price of the technology for a potential transaction is not a proper evaluation method. In fact, the variables used to populate the sheet are definitely key aspects when evaluating the potential of a novel product or process. In the chart below (Fig. 9.11), the user can have a graphical overview of the values assigned to the six major variables in this scenario. The selected variables are represented in the PVAE worksheet (Table 9.9).

In this third scenario it is easy to infer from the table above that the technology/patent at stake is not particularly important for the implementation of the final product/process, kind of weak from a legal standpoint, not that promising in terms of future development, very risky from a regulatory perspective, but destined to enter a viable market [59].

Once again, the PVAE worksheet also automatically generates useful graphs that allow the user to have a graphical perspective of the different scenarios. In the case at stake, the ggraphs generated by the worksheet are shown in Figs. 9.12–9.15.

Therefore, the PVAE method shows how it is possible to have a valuation of a technology or patent that also takes crucial,

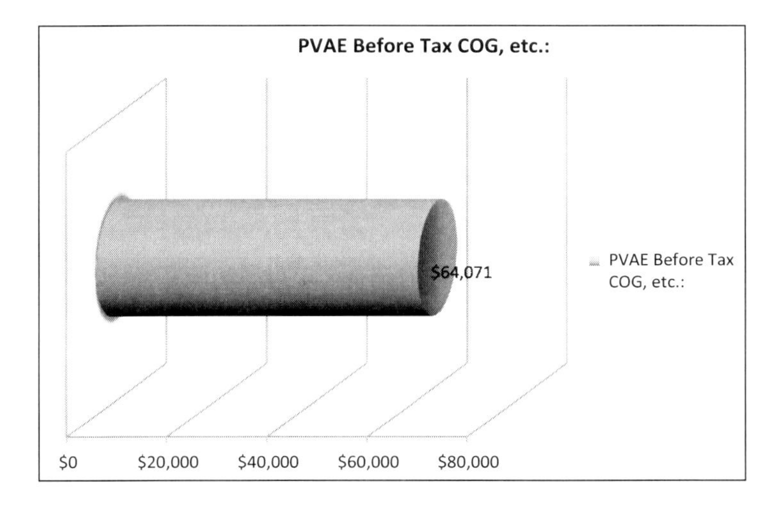

Figure 9.11 PVAE chart before tax, COG, etc.

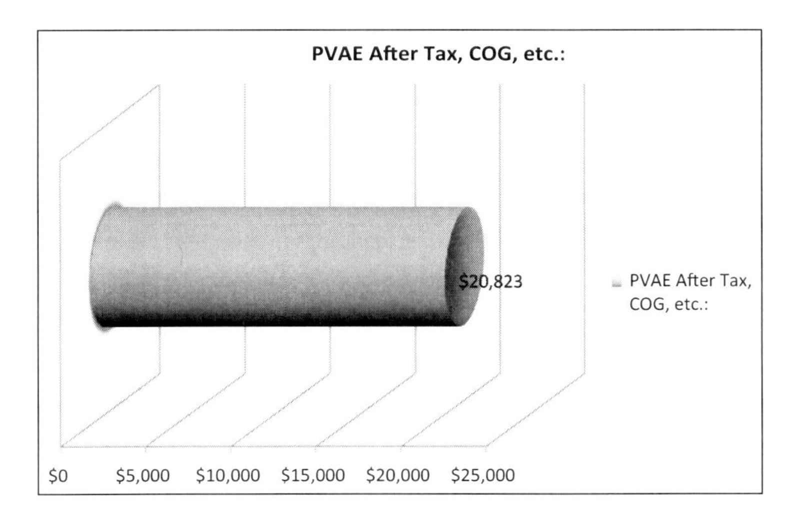

Figure 9.12 PVAE chart after tax, COG, etc.

Table 9.9 PVAE worksheet in the third scenario

Present value after evaluation							
Patent relevance 1–5:	2						
Initial cash flow:	$100,000			Technology 1–5	Financing 1–5	Regulation 1–5	Market 1–5
Patent coverage 1–5:	2			2	5	2	2
Years:	1–5	6–10	11–15				
Growth rate:	30%	20%	20%				
	Before tax, COG, etc.	After tax, COG, etc.					
Total of cash flows	$12,741,535	$4,140,999	Discount rate:	40%		Tax rate:	35%
Total PV of cash flows:	$728,080	$236,626	Royalty rate:	10%		COG, etc.	50%
Total PV of cash flows Considering the relevance of the patent and Patent coverage:	$116,493	$37,860	Partial PVAE	$46,597	$116,493	$46,597	$46,597
Present value of royalties (after tax)	$5,324		PVAE before tax COG, etc.:	$64,071	PVAE after tax, COG, etc.:	$20,823	
Year	Flows	Growth	Present value	Year	Rate	Royalty-tax	Total
1	$130,000	30%	$92,857	1	10%	$8,450	$8,450
2	$169,000	30%	$86,224	2	10%	$10,985	$19,435
3	$219,700	30%	$80,066	3	10%	$14,281	$33,716
4	$285,610	30%	$74,347	4	10%	$18,565	$52,280
5	$371,293	30%	$69,036	5	10%	$24,134	$76,414
6	$445,552	20%	$59,174	6	10%	$28,961	$105,375
7	$534,662	20%	$50,720	7	10%	$34,753	$140,128
8	$641,594	20%	$43,475	8	10%	$41,704	$181,832
9	$769,913	20%	$37,264	9	10%	$50,044	$231,876
10	$923,896	20%	$31,941	10	10%	$60,053	$291,929
11	$1,108,675	20%	$27,378	11	10%	$72,064	$363,993
12	$1,330,410	20%	$23,467	12	10%	$86,477	$450,470
13	$1,596,492	20%	$20,114	13	10%	$103,772	$554,242
14	$1,915,790	20%	$17,241	14	10%	$124,526	$678,768
15	$2,298,948	20%	$14,778	15	10%	$149,432	$828,200

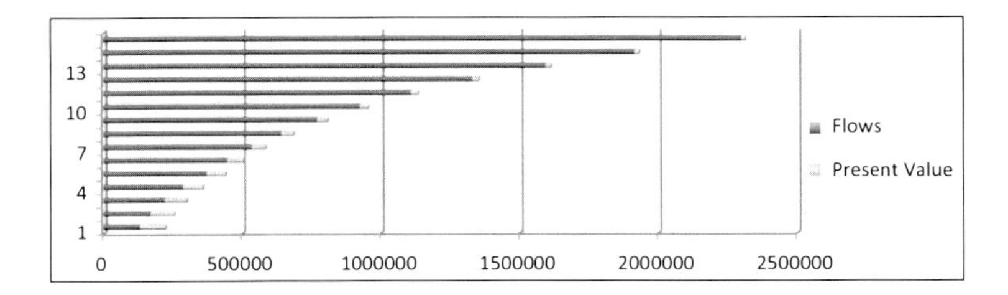

Figure 9.13 Future cash flows and present value.

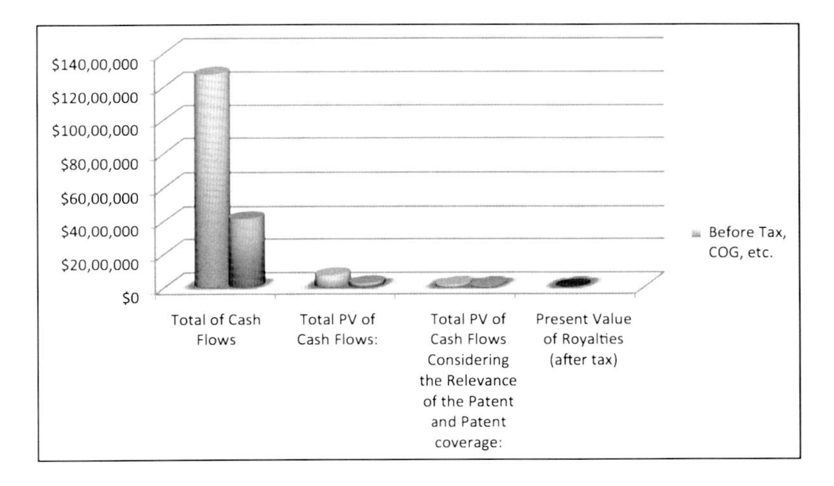

Figure 9.14 Cash flows and royalties.

nonquantitative variables into account, such as the environmental and/or human safety concerns that might adversely affect a technology's potentially brilliant future. These risks can be quite intense when dealing with nanotechnology-related products or processes. Due to the current rather unstable situation of this market, normative and regulatory barriers could pose serious commercialization barriers and as demonstrated above, by changing the qualitative variables the valuation may vary greatly. Indeed, the nature of nanotechnology at the moment can be further compromised by unexpected outcomes related to the very nature of the innovation in question, and these events may change the value of the product or process incorporating the technology, dramatically.

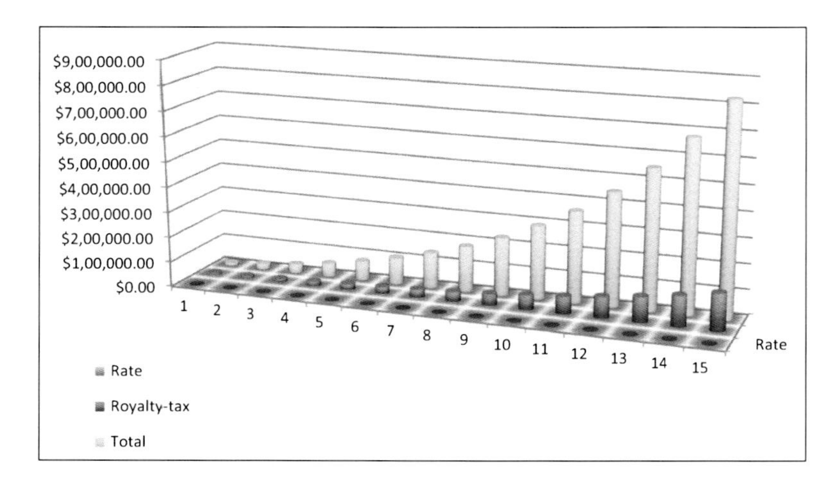

Figure 9.15 Royalty revenues.

In the realm of human safety and environmental preservation, these technologies could imply serious hazards, for example. Moreover, if a toxic substance is not totally metabolized in the human body, it can be excreted into the environment and endanger other humans as well. Unfortunately, if a product is marketed before all the required or suggested tests, the only way to see whether there are undesired effects is to wait. In this regard, the use of evaluation techniques and especially of the PVAE method to ponder all the variables involved in the commercialization of nanotechnology-based innovations would allow value adjustments, depending on potential changes to the situation. Therefore, licensing agreements could, for example, be linked to milestones or events that should allow the parties to change the variables and the values, accordingly.

9.4 Conclusions

Due to the dominant role played by universities and PROs in the conception and management of nanotech inventions, it is critical to devise a tool or a set of tools that could guarantee to the stakeholders that public research, performed with public money, has been appropriately valued before being licensed or sold to private or public companies. Nanotechnology, for its intrinsic peculiarities, seems to be the best candidate for testing the method and tool proposed in this chapter.

In this regard, the opportunities offered by the PVAE method and tool are quite remarkable. In fact, by using the total present value of the foreseen royalties and the PVAE, the user will come to two figures that can serve as a range to be used during negotiations. Also, there is no predetermined scenario when using the PVAE tool. If the technology has all top qualitative variables, then the PVAE will be greater than the present value of foreseen royalties (taking into account an average royalty rate). In the opposite case, when the qualitative variables are poor, it is very likely that the PVAE will be well below the present value of the expected royalties.

The University of Trieste (Italy) has officially started testing the PVAE tool in 2011 to provide all potential assignees/licensees of some selected technologies with a reliable method for the assessment of the value (post-evaluation) of its posted available technologies. The tool has been applied indistinctly to various technologies of the university, including those stemming from nanotech research. The PVAE seems to be especially effective in the university–industry setting, where companies usually take advantage of their stronger position and there is not yet a simple way of finding an objective range to use as a starting point for negotiations. The PVAE method constitutes therefore an interesting option to evaluate technologies or patents that pertain to particularly "risky" fields, where environmental and human safety issues might hinder a prosperous future for the technology or patent in question.

In fact, proper use of the PVAE tool would allow the parties to a license agreement or assignment to change the variables and adjust the value to the current situations. For instance, a novel molecule engineered through nanotechnology could face a hard time in government approval because of potentially adverse side effects to humans and/or the environment. In this case, the parties could simply adjust the technology and regulation variables to make the value of the innovation more adherent to its current market potential. It must be said that the PVAE tool, as any other valuation method, is purely subjective in two ways. First, it is necessary to have two parties agreeing to adopt a particular method rather than another. Once the tool or method is agreed upon, then the parties must find a common approach to play with variables and parameter ranges. Since the PVAE tool is based on projections and rates purely conceived by the parties, it is necessary in the first place that the parties agree to the numbers to be used in the worksheet and only then they will get an opportunity to assign a price value (and then a price) to a given technology by using as a reference the two values provided by the worksheet. In fact, what is necessary to comprehend, when the valuation of IP is concerned, is that while the cost is something undisputable as it is corroborated by factual findings, the price is determined by one party, and it is market efficient as long as the other party is willing to pay that given price. Finally, the value is the most difficult figure to agree upon as the interests of the parties are necessarily opposite. The seller shall assign a high value to the product to be sold and the buyer shall try to find or the possible arguments to show that the value is definitely inferior to the one asked for by the seller; therefore, the price should be found in between. The importance of the PVAE method, as said, is related to the fact that there are two values generated, and therefore, the parties who agree to use it know they will start negotiating by using the two figures as confines of the product's price latitude.

References

1. For a study based on the role of university spin-offs in nanotechnology, and their importance in technological change and further development, see Libaers, D., Meyer, M., Geuna, A. (2006) The role of university spinout companies in an emerging technology: the case of nanotechnology. *J Technol Transfer* **31**, 443–450. The authors highlight that the role of university spin-offs is definitely important, but there are some subfields in which these companies do not have a strong and/or growing proprietary technology.
2. Palmberg, C. (2008) The transfer and commercialization of nanotechnology: a comparative analysis of university and company researchers. *J Technol Transfer* **33**, 631–652.

3. Kesan, J. P. (2009) Transferring innovation. *Fordham Law Rev* **77**, 2169–2223.
4. Bastani, B., Mintarno, E., Akers, J., Fernandez, D. (2004) Technology transfer in nanotechnology: licensing intellectual property from universities to industry. *Nanotech Law Bus* **1**, 166–173.
5. Stewart, J. (2005) The nanotech university spinout company: strategies for licensing, developing, commercializing, and financing nanotechnology. *Nanotech Law Bus* **2**, 365–375.
6. For a good example of how technology services may be offered in the nanotech field, see Chiesa, V., De Massis, A., Frattini, F. (2007) How to sell technology services to innovators: evidence from nanotech Italian companies. *Eur J Innov Manage* **10**(4), 510–531. The study focuses on nanotech technical and scientific seminars companies.
7. Nanoforum (March 29, 2007) *Commercialization of Nanotechnology: Key Challenges*, http://www.nanoforum.org/dateien/temp/nano%20and%20security%20report%20June%2007.pdf?07022011203931.
8. Ibid
9. Ibid, 23–24.
10. Grace, R. H. 92007) *The 2007 Report Card on the Barriers to the Commercialization of Microelectromechanical Systems (MEMS) and Nanotechnology*, http://www.rgrace.com/documents/MEMSReportCard_2007.pdf.
11. Ibid, 5–6.
12. US Department of Commerce. (September 2007) *Barriers to Nanotechnology Commercialization: Final Report*, http://www.ntis.gov/pdf/Report-BarriersNanotechnologyCommercialization.pdf.
13. Ibid, 7.
14. Ibid
15. For an insightful article on entrepreneurship education, see Lehrer, M., Asakawa, K. (2004) Pushing scientists into the marketplace: promoting science and entrepreneurship. *Calif Manage Rev*, **46**(3), 55–76. The authors compare the experience of different countries as far as entrepreneurship among scientists is concerned and they suggest that institutions should start being entrepreneurial as well if the other countries are interested in catching up with the United States. They also conclude that the US leadership is not related to major skills or professionalism of its scientists but it is rather due to the competitive environment in which they have to operate. See also Johnson, D. (2006) Entrepreneurship education: towards a discipline-based framework. *J Manage Dev* **25**(1), 40–54.

16. Organisation for Economic Co-operation and Development. (2010) *The Impacts of Nanotechnology on Companies: Policy Insights from Case Studies*, OECD.
17. Ibid, 97.
18. In this regard the biotech era can teach something to entrepreneurs in the nanotech field. For a note on biotech business development, see Hamermesh, R. G., Higgin, R. F. (January 19, 2007) Note on biotech business development. *Harvard Bus Rev*, 1–24.
19. For an insightful analysis on the history of the commercialization of a magnetic resonance imaging (MRI) breakthrough, see Fleming, L., Aptekar, J. (September 26 2007) Commercializing an MRI breakthrough. *Harvard Bus Rev*, 1–14. See also Ruback, R. S. (March 25 2003) Merck & Company: evaluating a drug licensing opportunity. *Harvard Bus Rev*, 1–6.
20. For a general overview on valuation for venture capital, see Rosemberg, T. (March 23, 2009) A note on valuation for venture capital. *Harvard Bus Rev*, 1–14. See also Roberts, M. J. (October 24, 2005) Funding new ventures: valuation, financing, and capitalization tables. *Harvard Bus Rev*, 1–5.
21. Garrett, D. (2005) Break-out in nanotech: the next potential wave of IPOs. *Nanotech Law Bus* **2**, 274–277.
22. For an explanation of the electronic Dutch auction, see Rice, D. T. (2006) When the nanotech company goes public: using the electronic Dutch auction. *Nanotech Law Bus* **3**, 185–202.
23. Chin, P. (2005) Assessing venture capital returns for efficient investing in nanotechnology. *Nanotech Law Bus* **2**, 72–85.
24. Graffagnini, M. J. (2008) Raising venture capital for a nanoparticle therapeutics company. *Nanotech Law Bus* **5**, 207–216. In this paper Graffagnini explains why nanoparticle therapeutic companies usually require more financing and discusses the typical conditions of investments in these companies.
25. Munari, F., Toschi, L. (2007) How good are VCs at valuing technology? An analysis of patenting and venture capital investments in nanotechnology. *EPIP-2007 Conf. Mater*, 1–27.
26. Ibid, 16.
27. Ibid
28. Ibid
29. European Commission. (2004) *Towards a European Strategy for Nanotechnology* COM(2004)338, 12.5.2004.

30. Ibid, 16.
31. For further details on the European Parliament's opinion, see European Commission. (2005) *Nanosciences and Nanotechnologies: An Action Plan for Europe 2005–2009*, http://www.europarl.europa.eu/sides/getDoc.do;jsessionid=E9DBE5F6209124925273B577A43406F6.node1?language=EN&pubRef=-//EP//NONSGML+REPORT+A6-2006-0216+0+DOC+PDF+V0//EN.
32. Ibid, 5.
33. CNN Tech, http://articles.cnn.com/2011-03-08/tech/facebook.over-valued_1_mark-zuckerberg-facebook-worldwide-users?_s=PM:TECH.
34. Bismuth, A., Tojo, Y. (2008) Creating value form intellectual assets. *JIC* **9**(2), 228–245. This is an insightful paper in which the authors try to identify and discuss the use of intellectual assets in OECD countries.
35. Luehrman, T. A. (May 1, 1997) What's it worth? A general manager's guide to valuation. *Harvard Bus Rev*, 1–11. This is a very practical work showing how to compute the cost of capital for the discounted cash flow analysis.
36. Andriessen, D. G. (2005) *Value, Valuation and Valorisation*, http://www.openinnovatie.nl/downloads/Value_Valuation_and_Valorisation.pdf.
37. Ibid, 2–3.
38. Foster, G., Piotroski, J., Jia, N., Haemmig, M., Leslie, S. G., Tung. J. (February 5, 2009) Baidu.com, Inc.: valuation at IPO. *Harvard Bus Rev*, 1–28. See also Luehrman, T. A., Quinn, J. (April 19, 2010) Groupe Ariel S. A.: parity conditions and cross-border valuation (brief case). *Harvard Bus Rev*, 1–8.
39. Smith, G. V., Parr, R. L. (2000) *Valuation of Intellectual Property and Intangible Assets*, 3rd ed., John Wiley & Sons, New York. The work of Smith and Parr is the most comprehensive in terms of explained doctrines and case studies.
40. For a more synthetic overview of patent valuation practices, see Bauder, M. A., Rüether, F. (June 2009) Still a long way to value-based patent valuation: the patent valuation practices of Europe's top 500. *Les Nouvelles* http://www.wipo.int/edocs/mdocs/sme/en/wipo_insme_smes_ge_10/wipo_insme_smes_ge_10_ref_theme06_01.pdf. Also see Organisation for Economic Co-operation and Development. (June 30, 2005) *Intellectual Property as an Economic Asset: Key Issues in Valuation and Exploitation*, http://www.oecd.org/dataoecd/62/52/37031481.pdf. Bishop, J. C. (2003) The challenge of valuing intellectual property assets. *Nw J Technol Intell Prop* **1**(1) 59–65. See also Drews, D. (2004) *Intellectual Property Valuation Techniques*, http://www.ipmetrics.net/IPVT.pdf.

41. Flignor, P, Orozco, D. (June 2006) *Intangible Asset & Intellectual Property Valuation: A Multidisciplinary Perspective*, http://www.wipo.int/sme/en/documents/ip_valuation.htm.
42. It must be noted that several efforts have been made to adopt real valuation standards. See Cromley, J. T. (2007) Intellectual property valuation standards. *Intell Prop Today*, 36–38.
43. See also Chen, S. (2003) Valuing intellectual capital using game theory. *JIC* **4**(2), 191–201. The author suggests the combination of options theory with game theory to value intangible assets so that the value could adhere to market fluctuations and business strategies. See also Reuer, J. J., Tong, T. W. (2005) Real options in international joint ventures. *J Manage* **31**, 403–423.
44. Kossovsky, N. (2002) Fair value of intellectual property: an options-based valuation of nearly 8000 intellectual property assets. *JIC* **3**(1), 62–70. For an insightful article examining an options-based valuation of almost 8000 IP assets see Laxman, P. R., Aggarwal, S. (2003) Patent valuation using real options. *IIMB Manage Rev*, 44–51. This article presents a case study by adopting the real options approach on a PCT application. See also Wu, M. C., Tseng, C. Y. (2006) Valuation of patent: a real options perspective. *Appl Econ Lett* **13**, 313–318; Pitkethly, R. (1997) *The Valuation of Patents: A Review of Patent Valuation Methods with Consideration of Option-Based Methods and the Potential for Further Research*, Judge Institute Working Paper WP 21/97, 1–30. See also Sudarsanam, S., Sorwar, G., Marr, B. (2006) Real options and the impact of intellectual capital on corporate value. *JIC* **7**(3), 291–308.
45. Potter, R. H. (2007) Technology valuation: an introduction, in Krattiger, A., Mahoney, R. T., et al (eds.), *Intellectual Property Management in Health and Agricultural Innovation. A Handbook of Best Practices*, Oxford and Davis, MIHR and PIPRA.
46. Ibid, 809–810.
47. Kamiyama, S., Sheeahan, J., Martinez, C. (2006) *Valuation and Exploitation of Intellectual Property* STI Working Paper 2006/5, 1–48.
48. Ibid, 27.
49. Wise, R. M. *Valuation of Intellectual Property Assets: The Foundation for Risk Management and Financing*, http://www.wiseblackman.com/english/articles2.htm. This article presents a very profound analysis of the income approach and its subcategories, a must-read for all those interested in knowing the taxonomy of IP valuation.

50. Mard, M. J., Hiden, S., Rigby, J. S., Jr. (April 20, 2000) *Intellectual Property Valuation*, https://www.fairmarketvalue.com/page.php?content=buildingvalue. The article shows an interesting approach to determining the discount rate in the discounted cash flow approach.
51. Coleman, K. G. (2005) Nanotechnology company valuation. *Nanotech Law Bus* **2**, 376–382.
52. Ibid, 379.
53. Wartburg, I. V., Teichert, T. (2008) Valuing patents and licenses from a business strategy perspective: extending valuation considerations using the case of nanotechnology. *World Patent Info* **30**, 106–114.
54. For a general note on international licensing basics, see Beamish, P. W. (November 28, 2005). *Note on International Licensing*, 1–18, Ivey.
55. For a series of case studies on licensing in the semiconductor and electronics industries, see Grindley, P. C., Teece, D. J. (January 1 1997) Managing intellectual capital: licensing and cross-licensing in semiconductors and electronic. *Harvard Bus Rev*, 1–35.
56. The PVAE method has been introduced in two other prior publications in Escoffier, L. (2011) Reinterpreting patent valuation and evaluation: the tricky world of nanotechnology. *Eur J Risk Regul*, **1**, 67–78, and reprinted with permission in Escoffier, L. *Reinterpreting Patent Valuation and Evaluation: The Tricky World of Nanotechnology*, TTLF Working Paper No. 8, http://www.law.stanford.edu/program/centers/ttlf/papers/escoffier_wp8.pdf.
57. From January 2011, the University of Trieste has been the exclusive user of the PVAE tool to assign a value to its available technologies (http://www2.units.it/tto/). The PVAE method and its use have also been mentioned in Escoffier, L. (April 2011). *Innovating Innovation: Proposte per Migliorare il Trasferimento Tecnologico*, INNOV'AZIONE, 10–12.
58. For a detailed explanation of Porter's analysis, see Porter, M. E. (1979) How competitive forces shape strategy. *Harvard Bus Rev*, **57**(2), 137–145.
59. In this scenario, the technology should be deployed directly from the buyer; therefore, the financing variable equals 5.

Notes on the Contributors

Dr. Paris is a physician with remarkable volunteering experience in international settings, including India and Southeast Asia, where he has travelled repeatedly, from small villages to busy metropolitan areas, to provide health services to sick children. These activities as well as his professional interest in neurological disorders nudged him toward studying the impact of new technologies on improving the sensibility and sensitivity of current diagnostic tests and the efficacy of future therapeutic strategies. He is currently a fellow in the emergency department of St. Thomas' Hospital, London. He can be reached at Marco.Paris@gstt.nhs.uk.

Dr. Nicassio is a neurosurgeon with extensive experience, including previous working or honorary appointments in the National Hospital for Neurology and Neurosurgery (NHNN), London, and Addembrooke's Hospital, Cambridge; Johns Hopkins Hospital, Baltimore, and Barrow Neurological Institute (BNI), Phoenix; and continental Europe. His keen interest in microsurgical anatomy of the brain and spine led him to join forces with prominent colleagues, including Dr. Ganau, co-editor of this book, to provide young students and residents with an Italian edition of *Rhoton's Cranial Anatomy and Surgical Approaches*, one of the most detailed books unveiling the anatomy of the human central nervous system. Dr. Nicassio can be reached at Nicola.Nicassio@kch.nhs.uk.

Dr. Ligarotti graduated in medicine from the University of Milan (Italy), with a research thesis written in collaboration with the National Institute for Neurology "C. Besta." Currently he is completing his neurosurgical training as senior resident at the General Hospital "Niguarda" in Milan. Since 2003 he is in active service as medical officer of the Italian Air Force. Thus by working both in the neurosurgery department of the Military General Hospital "Celio" in Rome (Italy) as well as in several battlefields (Iraq and Afghanistan), he has forged a strong experience in neurotraumatology. His research activity led him to the International Research Base Camp "Concordia" (South Pole), where he spent the Antarctic summer of 2006–2007 (PNRA XXVII) and started developing his interest in the field of nano-innovation and its possible application in vascular and oncological neurosurgery. He can be reached at gianfranco.ligarotti@aeronautica.difesa.it.

Dr. Bosco graduated in material science and engineering from the University of Padua (Italy), working on elasticity and chirality in liquid crystals. He then defended his PhD in physics and chemistry of biological systems at the International School for Advanced Studies (SISSA) in Trieste (Italy) with a thesis on the elastic properties of nucleic acids, elucidated from both an experimental and a computational-theoretical perspective. During this period he forged a strong experience in computational simulations and bio-modeling of molecules and mastered experimental single-molecule techniques such as atomic force microscopy and optical tweezers. He currently holds a postdoc position in the NanoInnovation Lab at Elettra Synchrotron Radiation Facility in Trieste, supported by a grant from the Italian Association for Cancer Research (AIRC) and aimed at optimizing the fabrication of DNA-based biosensors for early-stage detection of markers involved in the development of tumors. He can be reached at alessandro.bosco@elettra.eu.

Dr. Parisse started his career as a physicist trained in surface physics and then obtained a PhD in physics from the University of L' Aquila (Italy), gaining further experience in the growth and characterization of molecular thin films for organic electronics applications, through both electron spectroscopy and scanning probe microscopy techniques. He subsequently moved his interests toward the self-assembling of biological molecules on surfaces, in crowded and confined environments, to elucidate their structure and functionality in living systems and to realize novel devices for in vitro proteomics. He is currently postdoctoral fellow in the

Elettra–Sincrotrone Trieste Unit of the National Interuniversity Consortium of Materials Science and Technology (INSTM Trieste Italy), where he is exploiting atomic force microscopy (AFM) and AFM-nanolithography to study enzymatic reactions on single DNA molecules and DNA nanostructures. Author of more than 20 papers in peer-reviewed international journals and several presentations at international conferences, he can be reached at pietro.parisse@elettra.eu.

Dr. Casalis, a physicist from the University of Pisa (Italy), with a PhD in condensed matter physics from the University of Trieste (Italy), is currently senior scientist at the Italian Synchrotron Radiation Facility, Elettra, where she leads the NanoInnovation Laboratory.

She started her postdoc tenure as a beamline scientist at Elettra, before being appointed as visiting scientist (2000–2002) at Princeton University (NJ, USA), where she worked in the group of Prof. Giacinto Scoles, using surface diffraction techniques to study thin films of organic molecules relevant to organic electronics applications. Back in Trieste, she focused on the application of atomic force microscopy to study interactions between biological molecules at surfaces and to design novel biosensors.

She sits in the board of experts of the PhD program in nanotechnology at the University of Trieste and of the PhD program in neurobiology at the International School for Advanced Studies (SISSA) in Trieste. She can be reached at loredana.casalis@elettra.trieste.it.

Dr. Israel Foroni graduated in theoretical nuclear physics from the University of Padua (Italy). After his doctoral degree he spent several years abroad (1982–1987) with highly prestigious scholarships at the Holt Radium Institute in Manchester and Hammersmith Hospital in London (U.K.), the Rambam HeatlhCare Campus in Haifa (Israel), and the National Cancer Institute in Bethesda (USA). Since 1988 he is working in the neurosurgical department of the University Hospital in Verona (Italy), where he cofounded the Minimally Invasive Neurosurgery Unit and became director of the Surgical Robotics Lab. In clinics he is primarily responsible for image processing and stereotactic treatment planning of deep-brain stimulation, gamma knife radiosurgery, and hyperthermia procedures, while his teaching and research activities as principal investigator in several international research consortia mainly revolve around intraoperative monitoring, vascular connectivity reconstruction, graphical models for awake neurosurgical procedures, prediction signals and models for comatose patients, microrobots, and nanopharmacology. He can be reached at foroni@bwh.harvard.edu.

Prof. Ambu graduated in medicine from the University of Cagliari (Italy), where he also pursued a specialization in pathology, with a subsequent fellowship in immunohistochemistry at the University of Leuven (Belgium). Back in Sardinia he started his clinical activity in the Institute of Pathology at the Department of Surgical Science of the San Giovanni Hospital in Cagliari, before being appointed to his first teaching position in the School of Medicine. Working in the group led by Prof. Gavino Faa, Prof. Ambu forged, along the years, a strong expertise in advanced diagnostic and prognostic techniques, deepening the understanding of both genetic and transcriptional bases of cancer and improving the phenotype classification of degenerative diseases and their neoplastic transformation. An international member of several scientific societies, Prof. Ambu currently sits in the board of experts of the PhD program in biomedical engineering at the University of Cagliari. He can be reached at amburo@unica.it.

Dr. Prisco graduated in medicine in 2007 and specialized in anesthetics, intensive care, and hyperbaric medicine in 2012 from the University of Trieste (Italy). She was appointed research fellow at the Coma Science Group in Liège (Belgium), before moving to U.K., where she started working as an anesthetist and intensivist at the Neurosurgical and General Intensive Care Unit of the University College Hospital in London and more recently at the Neurosciences Critical Care Unit of the Addenbrookes Hospital in Cambridge. A Singularity University alumna, she also participated in several highly competitive graduate programs related to space medicine (European Space Agency), brain plasticity (Neuroscience Department, Utrecht University), and international bioethics (Harvard University). In 2013 she obtained a master in science degree in pediatric intensive care medicine at Catholic

University in Rome (Italy). Chair-elect of the NEXT Committee of the European Society of Intensive Care Medicine, she can be reached at lara.prisco@singularityu.org.

Laura Ganau, although still a student of medicine at the University of Cagliari (Italy), has already brilliantly participated in several research activities mainly focused on neuroscience and related clinical fields. Her continuous efforts to play an active role in those projects, and to increase her promising amount of knowledge in such a challenging area of studies, allowed her to be among the recipients of important scholarships and grants from the Italian Ministry of Foreign Affairs, and even the US National Institute for Neurological Disorders and Stroke (Grant No. 1R13NS077709-01). A regular attendee of international scientific meetings (AANS/CNS, IBIA, AsMA, etc.), and a presenter of award-winning posters (i.e., Keystone Symposia in Molecular Biology), she has authored several articles published in peer-reviewed journals, as well as some book chapters. She can be reached at lolly26it@yahoo.it.

Sarah Rouse Janosik, PhD, JD, is senior counsel, intellectual property (IP) and litigation, at Amgen Inc. Dr. Rouse is a named inventor on various patents directed toward nanomedicine. Her research led to the formation of Keystone Nano, a company providing platform technologies for nanoenabled therapeutics. She received dual undergraduate degrees from the South Dakota School of Mines and Technology and her PhD in materials science and engineering from Pennsylvania State University (PSU). Her doctoral research focused on the synthesis, dispersion, and characterization of nanocomposite particles for bioimaging, drug delivery, and gene therapy. While at PSU, Dr. Rouse was named a National Science Foundation Fellow. She received her JD and certificate in IP from the DePaul University College of Law. She was the IP counsel at Onyx Pharmaceuticals until its acquisition by Amgen in 2013, and represented Onyx in the IP diligence in that transaction. Dr. Rouse practiced at law firms in the U.S. and Brazil and prior to that interned at the World Intellectual Property Organization (WIPO) Coordination Office at the United Nations.

Wim Helwegen holds a master of laws degree in international and European law from Tilburg University the Netherlands and a doctorate in commercial law from the University of Helsinki Finland. He is specialized in the interaction of patent law and advanced technologies, such as nanotechnology and biotechnology. After having served as a court clerk at a court of appeals, Helwegen wrote a doctoral dissertation on patents in the nanotechnology sector. Currently, he works as a legal advisor in IP and commercial contracts for law firms, companies and academic institutions.

Efrat Kasznik is an IP valuation and strategy expert with 20 years of consulting experience. She is the founder and president of Foresight Valuation Group, a Silicon Valley–based firm providing IP consulting and startup advisory services. She is also a lecturer on IP management at the Stanford Graduate School of Business. Kasznik specializes in analyzing IP for a range of purposes, including mergers and acquisitions, financial reporting, technology commercialization, transfer pricing, and litigation damages. She is listed on the IAM 300 list of leading IP strategists and is a member of the leadership committee of the Licensing Executives Society (LES) U.S.-Canada, High Tech Sector. Kasznik has been involved as a CFO, cofounder, and adviser to several start-ups and investment funds in the U.S. and Europe. She holds an MBA from UC Berkeley, Haas School of Business, and a BA in accounting and economics from Hebrew University, Jerusalem.

Index

Abraxane 93
Accounting Standards Codification (ASC) 153–154, 158, 161
AFM, *see* atomic force microscopy
AFM probes 39
antibiotic resistance 72
antibodies 32–36, 39–41, 83
antigens 34–35, 39, 41, 94–95
artificial cells 71
artificial nanomotors 70
ASC, *see* Accounting Standards Codification
assay 23, 33–35, 37
  sandwich 35–36
assets, nanotechnology-related 184
atomic force microscopy (AFM) 24, 30, 39, 62
Aurimmune 92

β-TCP, *see* β-tricalcium phosphate
β-tricalcium phosphate (β-TCP) 63–64
bacteria 72
Bayh–Dole Act 164
Berne Convention 145–146
biological barriers 9, 81, 83, 93
biological interfaces 74
biological molecules 67
biological motors 70
biomaterials 60
biomolecules 12, 27–29, 32, 45
biosensors 5, 22, 24, 30, 33–35, 37–39, 47–48, 59
biotechnological applications 61
  consumer 23
biotechnology 26, 93, 165, 177
block copolymer micelles 88
blood 14, 62, 71–72
blood vessels 48, 71, 82–83
blood–brain barrier 9, 82–83
bone formation 63
bones 63
brain 14, 74
brain–computer interfaces 74–75
brain–machine interfaces 73

cancer 21–22, 42, 49, 51, 81–82, 90, 93
  stomach 86, 88
cancer cell imaging 44
cancer cells 44–45, 50, 84, 92
cancer therapy 12
cancer treatment 90, 93
carbon dioxide 71–72
carbon nanotubes (CNTs) 44–45, 66, 74
cells
  healthy 92–93
  single 7, 43, 47, 68
  stem 50, 185
circulating tumor cells (CTCs) 50
CNTs, *see* carbon nanotubes
combined nanomedicine treatments 99

consumer price index (CPI) 181
coronary stenting 64
cost-sharing agreements 160
CPI, *see* consumer price index
CTCs, *see* circulating tumor cells

DDI, *see* DNA-directed immobilization
dendrimers 10, 84, 89
devices
  implantable 12, 46–47, 74
  microtechnological 24
diagnostics
  clinical 7, 42
  in vitro 41–45
  in vivo 23, 41, 46–47
dip-pen nanolithography (DPN) 24, 29–30
disclosure 138, 140
diseases 3, 5–6, 11, 16, 21, 23, 42, 75, 81, 111
DNA 22–23, 32–33, 40, 44, 70
DNA-directed immobilization (DDI) 34, 40–41
DNA microarray technology 40
doxorubicin 87–88, 90
DPN, *see* dip-pen nanolithography
drug delivery 10–11, 23, 48, 89, 92, 99, 112
drug resistance 84

EGFR, *see* epidermal growth factor receptor
electron-beam lithography 25–26
ELISA, *see* enzyme-linked immunosorbent assay
ELISPOT, *see* enzyme-linked immunospot assay
emulsions 85–86, 88
enzyme-linked immunosorbent assay (ELISA) 8, 34–35, 41
enzyme-linked immunospot assay (ELISPOT) 35–36
enzymes 33–34, 40, 43, 71–72
epidermal growth factor receptor (EGFR) 43, 50
epidural fibrosis 65–66
ethics 111–112, 114, 116–117

fair market value (FMV) 151, 159, 179
FMV, *see* fair market value

GAAP, *see* generally accepted accounting principles
generally accepted accounting principles (GAAP) 153
genes 49, 65, 71, 90
genetech 112–113
glucose 44, 47, 71–72
gold nanoparticles 91–92
gold nanorods 92
gold nanoshells 91–92
growth factors 12–13

health 98, 111, 122, 125, 172–173
healthcare spending 4
healthcare systems 4
HIV
  see human immunodeficiency virus
  transmission of 89
human immunodeficiency virus (HIV) 86, 89

IAM, *see* intellectual asset management
immune responses 94
immunosignal 39
implants 14, 75
infectious diseases 81–82

influenza 86
innovations
 nanotechnology-based 183, 204
 nanotechnology-related 170
intellectual asset management (IAM) 152, 163
intellectual property (IP) 115, 119, 136, 142, 144–145, 147, 151–152, 155–160, 162–165, 172, 174–175, 181, 188, 206
intellectual property evaluation 165–166
intellectual property rights (IPRs) 135–136, 138, 140, 142, 144, 146, 148–149, 175
intellectual property valuation 151
intellectual property valuation 151–152, 153–165, 206
invention, nanotechnology-related 186
IP, *see* intellectual property
IPRs, *see* intellectual property rights

joint ventures 164–165

leishmaniasis 82, 95
life expectancy 185, 191, 194, 199
Lipo-Dox 87
liposomes 9–10, 12, 84, 86–87, 126
lithographic techniques 24–25, 27
lithography, soft 26–27
litigation 142, 151–152, 155, 165–166, 179
litigation damages 155, 160

magnetic nanoparticles 93
magnetic resonance imaging (MRI) 23, 45–46, 48–49
mass transport 38
medical products 124, 126, 128
medication tolerance 92–93
medicinal products 126
medicine 4–6, 14, 16, 21, 75, 111, 128
metastases 47, 50
micelles 9–10, 84, 86, 88
microbivores 71–73
microchips 74
microcontact printing 24, 27–28
microelectromechanical systems 75
microfabrication 25–26
microfluidics 23, 27, 45
micronanoprojection vaccine delivery 95
miniaturization 22–24, 40–42, 47
MRI, *see* magnetic resonance imaging

nanobead vaccines 95
nanobeads 95
nanocomposites 62
nanocontact printing 27, 29
nanodevices 41, 66, 71, 74, 120
nanodrugs 10–12, 114
nanoemulsions 9, 11, 85–86
nanoenabled therapeutics 120, 123, 126–128
nanoethics 111–112, 114, 116
nanoformulations 82, 84–85, 87, 89, 91, 93, 95
nanofuture 111–112
nanogels 88–89
nanografting 24, 30–31
nanoknives 66–67
nanomedical-based products 120–121
nanomedicine, regulation of 124–127

nanomedicine patents 97
nanomedicine products 97–99
nanomedicine publications 96
nanomedicine research 16–17
nanoparticle suspensions 120–121
nanoparticles
  chitosan 95
  functionalized 48, 85
nanopores 9
nanorobotics 69, 71
nanorobots 5, 69–70
nanoshells 91–92
nanosurgery 66, 68–69
nanotech 170, 174–175
nanotechnological evolution 115
nanotechnologists 60
nanotechnology
  application of 15, 17, 128
  commercialization of 174–176
  molecular 112
nanotechnology-based paint 183
nanotechnology-based products 119, 126
nanotechnology commercialization 169–170, 172, 176
nanotechnology concepts 123
nanotechnology-derived medical solutions 6
nanotechnology development 170
nanotechnology innovations, commercialization of 169–175
nanotechnology IP holders 163
nanotechnology IP portfolio 163
nanotechnology patents 156
nanotechnology research 119, 125, 177
nanotechnology standardization 123, 128
nanotechnology standards 173
nanotechnology valorization 175
nanotherapeutics 81–84, 86, 88, 90, 92–94, 96–99
nanotweezers 66
nanovaccines 81–82, 95–96
navigating patent thickets 165
net present value (NPV) 153
neurosciences 75
neurosurgery 74
neurosurgical procedures 65
NPV, *see* net present value

optical tweezers 66–67
oxygen 71–72

paclitaxel 88, 90, 93
patent aggregators 155
patent donations 151
patent evaluation 176, 184
patent infringement 143, 151, 155
patent infringement damages 156–157
patent infringement litigations 156
patent lawsuits 140
patent litigation 155–156, 165
patent pooling 184
patent protection 149
patent relevance 187–188, 191, 196, 200
Patent Term Extension 136
patent thickets, resolving 165
patentability 137–138
patents
  nanotech 184
  pharmaceutical 136
pathology, removal of 73
PCU, *see* poly(carbonate-urea)urethane
PEG, *see* polyethylene glycol
PEGylation 90
peptides 10, 32, 34
PET, *see* positron emission tomography
pharmacokinetics 83–84, 88

photodamage 67
photodiodes 30–31
photolithography 24–25, 27
photomask 25–26
photosensitizer pyropheophorbide a (PPa) 44, 158
platelet adhesion 61
poly(carbonate-urea)urethane (PCU) 61
polyethylene glycol (PEG) 10–11, 90
polyhedral oligomeric silsesquioxane (POSS) 60–61
polyhedral oligomeric silsesquioxane-poly(carbonate-urea) urethane (POSS-PCU) 61
polymer conjugates 90
polymeric nanoparticles 90–91
polymers 11, 28, 60, 84, 88, 90
  synthetic 13
positron emission tomography (PET) 49
POSS, *see* polyhedral oligomeric silsesquioxane
POSS-PCU, *see* polyhedral oligomeric silsesquioxane-poly(carbonate-urea) urethane
PPa, *see* photosensitizer pyropheophorbide a
present value after evaluation (PVAE) 187–192, 194–196, 198–201, 205
proteins 8, 10, 13, 22, 27–28, 33–36, 40, 42–43, 67, 70–71, 90, 99
proteomics 22, 33, 36, 40
PVAE, *see* present value after evaluation

quantum computers 75
quantum dot nanocrystals 61

radiolabeled readout 34, 36–37
radiolabels 36
regenerative medicine 60–61, 63, 65

safety, human 186, 203–204
SAMs, *see* self-assembled monolayers
self-assembled monolayers (SAMs) 27, 31–32, 41
semiconductor industry 165
sensors 14, 37, 46–47, 92
silicon nanotweezers (SNTs) 67–68
single-cell analyses 42–43
single-walled carbon nanotubes (SWCNTs) 44
SNTs, *see* silicon nanotweezers
solid tumors 11, 87
stent thrombosis, late 64–65
stents, drug-eluting 64
surgical microscopes 69
SWCNTs, *see* single-walled carbon nanotubes

technology-based intangible assets 154
technology factor method 180
technology life cycles 174
technology risk of development 162
technology transfer 163–164, 176
TEM, *see* transmission electron microscopy
therapeutic agents 10–11, 48
trade secrecy 148–150
transhumanism 115–116

transmission electron microscopy
(TEM) 45
tuberculosis 82, 89, 94
tumor cells, circulating 50
tumors 11, 43–44, 48–49, 83, 89,
91–93
primary 50

US patents 139, 141

vasculoid 71
VCs, *see* venture capitalists
venture capitalists (VCs) 172,
175